MOLECULAR BIOLOGY
INTELLIGENCE
UNIT

Endosomes

Ivan Dikic, M.D., Ph.D.

Institute of Biochemistry II
Goethe University School of Medicine
University Hospital
Frankfurt, Germany

LANDES BIOSCIENCE / EUREKAH.COM
GEORGETOWN, TEXAS
U.S.A.

SPRINGER SCIENCE+BUSINESS MEDIA
NEW YORK, NEW YORK
U.S.A.

ENDOSOMES
Molecular Biology Intelligence Unit

Landes Bioscience / Eurekah.com
Springer Science+Business Media, LLC

ISBN: 0-387-39950-X   Printed on acid-free paper.

Springer Science+Business Media, LLC, 233 Spring Street, New York, New York 10013, U.S.A.
http://www.springer.com

Please address all inquiries to the Publishers:
Landes Bioscience / Eurekah.com, 810 South Church Street, Georgetown, Texas 78626, U.S.A.
Phone: 512/ 863 7762; FAX: 512/ 863 0081
http://www.eurekah.com
http://www.landesbioscience.com

Printed in the United States of America.

9 8 7 6 5 4 3 2 1

Library of Congress Cataloging-in-Publication Data

Endosomes / [edited by] Ivan Dikic.
    p. ; cm. -- (Molecular biology intelligence unit)
  Includes bibliographical references and index.
  ISBN 0-387-39950-X
1. Endosomes. I. Dikic, Ivan. II. Series: Molecular biology intelligence unit (Unnumbered)
[DNLM: 1. Endosomes--physiology. 2. Cell Physiology. QU 350 E56 2006]
QH603.E63E53 2006
571.6'55--dc22

2006018965

About the Editor

IVAN DIKIC received his M.D. and Ph.D. from the University of Zagreb, Croatia. After a postdoctoral tenure with Joseph Schlessinger at NYU, New York, USA, he became a group leader at the LICR, Uppsala, Sweden. In 2003, he was appointed a Professor of Biochemistry at the Goethe University Medical School, Frankfurt, Germany and from 2005 he co-ordinates work in the Tumor Biology Laboratory at the MedILS, Split, Croatia. His group focuses on the role of the E3 ligase Cbl and its associated proteins in signalling and receptor endocytosis and on the emerging role of ubiquitin and Ub-like modifiers as signalling devices controlling intracellular trafficking, gene transcription and DNA repair. He recently received 26th AACR Award for the Outstanding Achievement in Cancer Research.

CONTENTS

EDITOR

Ivan Dikic, M.D., Ph.D.
Institute of Biochemistry II
Goethe University School of Medicine
University Hospital
Frankfurt, Germany
Email: ivan.dikic@biochem2.de

CONTRIBUTORS

Dorothea Brandhorst
Department of Neurobiology
Max-Planck-Institute
　for Biophysical Chemistry
Göttingen, Germany
Email: dorothea.brandhorst@gmx.de
Chapter 5

Julien Chevallier
Department of Biochemistry
Sciences II
Geneva, Switzerland
Chapter 2

Silvia Giordano
Division of Molecular Oncology
University of Turin Medical School
Candiolo, Turin, Italy
Email: silvia.giordano@ircc.it
Chapter 8

F. Gisou van der Goot
Department Microbiology
　and Molecular Medicine
University of Geneva
Geneva, Switzerland
Email:
　gisou.vandergoot@medecine.unige.ch
Chapter 12

Jean Gruenberg
Department of Biochemistry
Sciences II
Geneva, Switzerland
Email: jean.gruenberg@biochem.unige.ch
Chapter 2

Gal Gur
Department of Biological Regulation
The Weizmann Institute of Science
Rehovot, Israel
Email: gal_gu@rosettagenomics.com
Chapter 9

Volker Haucke
Institut für Chemie-Biochemie,
Freie Universität Berlin
Berlin, Germany
Email: vhaucke@chemie.fu-berlin.de
Chapter 4

Reinhard Jahn
Department of Neurobiology
Max-Planck-Institute for Biophysical
　Chemistry
Göttingen, Germany
Email: rjahn@gwdg.de
Chapter 5

Letizia Lanzetti
Division of Molecular Angiogenesis
University of Turin Medical School
Turin, Italy
letizia.lanzetti@ircc.it
Chapter 8

Ira Mellman
Department of Cell Biology
Ludwig Institute for Cancer Research
Yale University School of Medicine
New Haven, Connecticut, U.S.A.
Email: ira.mellman@yale.edu
Chapter 1

Mark Marsh
Cell Biology Unit
MRC-Laboratory for Molecular
   Cell Biology
and
Department of Biochemistry
   and Molecular Biology
University College London
London, UK
Email: m.marsh@ucl.ac.uk
Chapter 11

Marta Miaczynska
International Institute of Molecular
   and Cell Biology
Warsaw, Poland
Email: miaczynska@iimcb.gov.pl
Chapter 3

Krupa Pattni
Department of Biochemistry
The Norwegian Radium Hospital
Montebello, Norway
Chapter 7

Simona Polo
IFOM, Istituto FIRC
   di Oncologia Molecolare
European Institute of Oncology
Institute for Cancer Research
   and Treatment (IRCC)
Milan, Italy
Chapter 8

Rosa Puertollano
Laboratory of Cell Signaling
National Heart, Lung,
   and Blood Institute
National Institutes of Health
Bethesda, Maryland, U.S.A.
Email: puertolr@mail.nih.gov
Chapter 10

Núria Reig
Department Microbiology
   and Molecular Medicine
University of Geneva
Geneva, Switzerland
Chapter 12

Oleg Shupliakov
Center of Excellence
   in Developmental Biology
Department of Neuroscience
Karolinska Institute
Stockholm, Sweden
Email: oleg.shupliakov@neuro.ki.se
Chapter 4

Harald Stenmark
Department of Biochemistry
The Norwegian Radium Hospital
Montebello, Norway
Email: stenmark@ulrik.uio.no
Chapter 7

Linton M. Traub
Department of Cell Biology
   and Physiology
University of Pittsburgh School
   of Medicine
Email: traub@pitt.edu
Chapter 6

Yosef Yarden
Department of Biological Regulation
The Weizmann Institute of Science
Rehovot, Israel
Yosef.yarden@weizmann.ac.il
Chapter 9

Marino Zerial
Max Planck Institute of Molecular Cell
   Biology and Genetics
Pfotenhauerstrasse
Dresden, Germany
Email: zerial@mpi-cbg.de
Chapter 3

Yaara Zwang
Department of Biological Regulation
The Weizmann Institute of Science
Rehovot, Israel
Email: yaara.zwang@weizmann.ac.il
Chapter 9

PREFACE

Endosomes are a heterogeneous population of endocytic vesicles and tubules that have captivated the interest of biologists for many years, partly due to their important cellular functions and partly due to their intriguing nature and dynamics. Endosomes represent a fascinating interconnected network of thousands of vesicles that transport various cargoes, mainly proteins and lipids, to distant cellular destinations. How endosomes function, what coordinates the molecular determinants at each step of their dynamic life cycle and what their biological and medical relevance is, are among the questions addressed in this book.

The past two decades have witnessed rapid strides in our understanding of the morphology and functions of endosomes (Chapter 1). In retrospect, the classical view of endocytic organelles has to give way to a more complex one, in that multiple, functionally distinct microdomains coexist within one endosomal structure. These microdomains are determined by a certain composition of distinct proteins or lipids that act as organizers of specific membrane domains (Chapters 2 and 3). Among the best-known facets of endosome function is their role in trafficking events at synapses, both in presynaptic and postsynaptic compartments of nerve cells (Chapter 4).

A detailed understanding of processes that regulate endosome fusion, clathrin-dependent receptor endocytosis and sorting to the recycling route or the degradative pathway is available via the integration of structural, molecular and biochemical studies on distinct endocytic adaptor proteins (Chapters 5, 6 and 7). Another important aspect is the processing of signals originating from internalized receptors and how their fate and signalling potency in cells are linked to their accumulation in distinct endosomal compartments, specifically during endocytosis of receptor tyrosine kinases (Chapters 8 and 9).

A topic of particular interest for the general public deals with the interface between endocytic trafficking and human diseases. Molecular views on aberrations in endosomes that are linked to the pathogenesis of various diseases are summarized (Chapter 10). The last two chapters are dedicated to the role of endosomes in viral entry and replication and how external pathogens and toxins hijack the endocytic machinery for their purposes by exploiting the cell's transport infrastructure (Chapters 11 and 12). Although quite similar in the general form of action many toxins appear to utilize different strategies to enter the cellular endosome system.

The concise format of the chapters and up-to-date molecular explanations of endosome functions may have a broad appeal for both students and scientists who wish to know more about the exciting world of trafficking in the cell.

Ivan Dikic

CHAPTER 1

Endosomes Come of Age

Ira Mellman*

"Endosomes are a population of endocytic vacuoles through which molecules internalized during endocytosis pass en route to lysosomes. In addition to this transport function, recent studies indicate that these organelles also act as clearing houses for incoming ligands, fluid components, and receptors."
—Helenius A et al. *Trends in Biochem Sci* 1983; 8:245-250.

In the late 1970s, a confluence of exciting observations triggered unprecedented interest in the functions and mechanisms of endocytosis. This activity, in turn, rapidly led to the identification of endosomes as a new and distinct organelle. Endosomes were first defined in 1983 by a simple statement that remains largely true even today (see above).[1] Endosomes were appreciated not only as intermediates on the pathway to lysosomes, but also on the pathway of receptor recycling where they were seen as being the primary site for the dissociation of ligand-receptor complexes and the return of unoccupied receptors back to the plasma membrane. In addition, for those receptors subject to "down regulation", endosomes were understood to be the place at which the crucial sorting decision was made between recycling and lysosomal transport. Many of these features were linked to the ability of endosomes to lower their internal pH via the activity of an ATP-dependent proton pump. Acidification was a key conceptual advance since it explained why many receptors discharged their ligands upon reaching endosomes and how incoming enveloped viruses fused with the endosomal membrane, initiating infection.

All this happened more than 21 years ago, meaning that even in Puritanical Western countries such as the United States, endosomes have (legally speaking) come of age. In other words, they can drink legally. This is a good thing since the primary activity mediated by endosomes is that of pinocytosis, literally "cell drinking".

Scientifically speaking, endosomes probably came of age long ago. Within a few years of their description, they became widely accepted as a new if incompletely understood organelle by cell biologists, with the concept rapidly spreading to allied fields. It is rare for a new organelle to be so quickly and broadly adopted, falling into the scientific lexicon so that endosomes now appear to have always existed (which, of course, they have). Yet, perhaps as a testament to their relative conceptual youth, our understanding of endosomes has continued develop at a remarkable pace. As will become abundantly clear in subsequent chapters, we have learned an immense amount about the mechanisms by which endosomes conduct their activities. Conversely, the study of endosomes has enabled the discovery of a wide array of basic principles in the broader field of molecular cell biology. This section will not attempt to review all of what we know concerning how endosomes work. It will, instead, take the opportunity to chart the development of how endosomes came to be understood, both functionally and mechanistically. The section is also written from the perspective of one who was privileged to be among

*Ira Mellman—Department of Cell Biology, Ludwig Institute for Cancer Research, Yale University School of Medicine, 333 Cedar Street, P.O. Box 208002 New Haven, Connecticut 06520-800, U.S.A. Email: ira.mellman@yale.edu

Endosomes, edited by Ivan Dikic. ©2006 Landes Bioscience and Springer Science+Business Media.

those present at the beginning. Other perspectives exist, and others have contributed to the overall development of the field, even if they have not all been highlighted in this brief chapter. The goal is to place current advances in the context of the relatively brief history of endosomes as central players in cell biology.

The Discovery of Endosomes

To place our current understanding of endosomes in context, it is important to understand the scientific background to their discovery. Without attempting to provide a complete or systematic account, a few key conceptual highlights bear mentioning.

Two key observations were made in the mid- to late-1970s that not only launched the modern field of endocytosis, but also put in motion the events that would eventually lead to the identification of endosomes as distinct organelles. The first was from Steinman and colleagues, who used quantitative biochemical and EM approaches to demonstrate that mammalian cells in culture (fibroblasts, macrophages) internalized enormous areas of plasma membrane during constitutive endocytic activity, 50-200% every hour.[2,3] They reasoned that this rate of internalization was far greater than the capacity of cells for de novo membrane synthesis and concluded that the bulk of membrane must be recycled intact back to the plasma membrane. The second was from Brown, Goldstein and colleagues who were characterizing the receptor-mediated endocytosis of LDL. Highly influential, these studies demonstrated (among many other things) that more LDL particles were taken up and degraded than could be accounted for by new receptor synthesis, and that uptake was initiated at clathrin-coated pits.[4-6] In other words, internalized receptors were selectively internalized and rapidly recycled.

The principle of recycling during ligand uptake was rapidly extended to many other types of receptors. Moreover, our own early work demonstrated that, in general, a wide variety of membrane proteins were susceptible to internalization even during constitutive endocytosis.[7] In all cases, the internalized pools of membrane proteins remained predominantly long lived ($t_{1/2}$ ~24 hr), and could thus inferred (or in some cases shown directly) to escape intracellular degradation and recycle to the surface multiple times.[8]

There was a major conceptual problem posed by recycling, however. At the time, all endocytosis was viewed as having lysosomes as the primary intracellular destination. Indeed, even in the case of LDL receptor, bound LDL was seen as being released from its receptor after lysosomal delivery; the rapidity of transit through the degradative compartment presumably facilitating the receptor's escape. Since it appeared unlikely that receptors could survive repeated exposures the lysosomal proteases, increasing attention began to be paid towards a poorly described set of structures variably referred to as pinocytic vesicles, endocytic vesicles, pinosomes, receptosomes, CURL, or (perhaps most commonly) prelysosomal vacuoles. First hinted at in early cytochemical studies of endocytosis in the kidney by Strauss, these structures generally appeared more phase and electron lucent than did hydrolase-rich lysosomes.[1] It rapidly became clear, however, that they behaved as intermediates on the lysosomal pathway, accumulating internalized tracers (fluid phase components, receptor-bound ligands) transiently and prior to lysosomal arrival. Importantly, they also were relatively low with respect to their content of hydrolytic enzymes.

The first real indication that these prelysosomal structures were more than intermediates emanated from evidence in intact cells that they were acidic, and therefore might have an intrinsically important function. Previously, only lysosomes were recognized as acidic organelles in most cell types. The work of Helenius and colleagues was especially important in this regard. In the course of studying the entry and infection of enveloped animal viruses, they found that the low pH-induced fusion event required for entry occurred kinetically well before delivery of virions to hydrolase-rich lysosomes.[9] Work from Maxfield and from Klausner involved exposing cells to ligands coupled to the pH sensitive fluorochrome, showing (either by fluorescence microscopy or fluorometry) that the probes reached vesicles of acidic pH shortly after entry, presumably prior to lysosomal arrival.[10,11]

Our work made use of the recently established fact that prelysosomal vacuoles had a density on Percoll gradients that was much lower than lysosomes. This enabled us to physically separate lysosomes from endosomes in cells exposed to pH probes. In vitro, the low density endosomes were capable of ATP-dependent acidification, indicating that they, like lysosomes, contained an ATP-driven proton pump.[12] Mutant cells defective in virus entry (and killing by pH-activated bacterial toxins) in vivo exhibited a selective defect in endosomal acidification in vitro.[12] It is important to emphasize that the Percoll gradient experiments demonstrated that these prelysosomal structures were acidic but did not have the abundant hydrolytic activity found in lysosomes. In other words, endosomes were physically and functionally distinct from lysosomes.

So rapid was the progress that the published record was woefully unable to keep pace with the new ideas and experiments coming from many different laboratories. All involved seemed to understand that a new organelle was being born and, even more importantly, that a fundamental new pathway was being defined. Even while the acidification story was being developed, groups working on receptor-mediated endocytosis began accumulating evidence that the acidic pH in endosomes was responsible for the dissociation of many receptor-ligand complexes—and before their delivery to lysosomes. Favorite experiments at the time involved treating cells with acidophilic weak bases (chloroquine, ammonium chloride) or carboxylic ionophores (monensin, nigericin) that dissipated endosome (and lysosome) pH, showing that ligand discharge and sometimes receptor recycling could be blocked. Similarly, incubation of cells at intermediate temperatures (<20°C) appeared to block transit of all internalized substances to lysosomes, but allowed for continued ligand uptake and receptor recycling.

More direct evidence came from the work of Geuze and colleagues who used the emerging technique of immuno-gold EM on ultrathin cryosections (combined with some biochemistry) to define a prelysosomal compartment in which receptor and ligand physically dissociated.[14] The acid-dependent discharge of iron from internalized transferrin (Tfn) was shown, in Percoll gradients, to occur in low density endosome-containing fractions.[10] Interestingly, Tfn itself was never found to transfer to high density fractions, suggesting that it recycled to and from endosomes without encountering lysosomes. Much the same was found for the FcγRII-B2 IgG receptor, which was shown to recycle from endosomes while monovalent, but transferred from endosomes to lysosomes when cross-linked.[15,16]

Such experiments provided powerful functional evidence that endosomes existed, and that they played an essential role in receptor recycling. As in biochemistry and murder investigations, however, unless an actual protein—or "body"—can be purified or produced, the existence of endosomes as a discrete compartment would remain in doubt. A great effort was thus expended attempting to isolate these structures. Cell fractionation is difficult under the best of circumstances, but a particular challenge was presented by endosomes, an organelle with no known intrinsic markers, with no obvious morphological features, and that was not particularly abundant. We were fortunate to obtain perhaps the most enriched endosome populations, using an electrophoretic technique that separated membranes on the basis of inherent net charge.[17,18] The resulting membrane fractions had a predictable composition, similar to but distinct from the plasma membrane and lysosomes. Endosomes exist.

Taking all this together, the view emerged that endosomes provided a distinct intracellular site at which internalized receptor-ligand complexes were first dissociated and then sorted to distinct destinations: receptors to the plasma membrane for reuse and discharged ligands to lysosomes for degradation.[1] Figure 1 presents a very early diagram of the pathway, which turned out to be fairly prescient in several respects.

The ability specify different fates for membrane and luminal components was remarkable, and further defined the ability of endosomes to sort the container from the fluid it contained. From both EM and quantitative measurements, it was concluded that a good fraction of this sorting activity could reflect simple Euclidean considerations. If endocytosis was mediated by spherical clathrin coated vesicles while recycling was initiated by the long tubular extensions

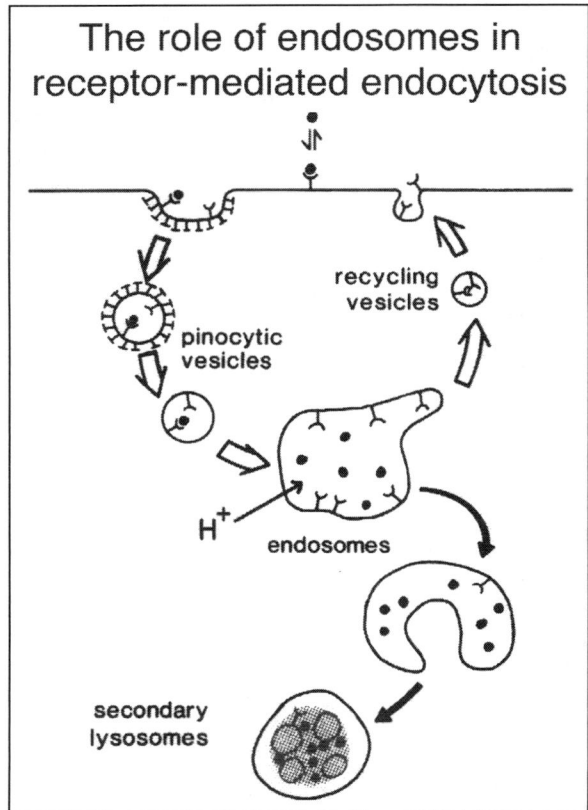

Figure 1. An early diagram of the role of endosomes in receptor-mediated endocytosis, circa 1983. Drawn by Ari Helenius, Ira Mellman, and Margaret Moench.

characteristic of endosomes, the surface to volume ratio of the recycling vehicles would be far higher than that of the incoming vesicles. Thus, recycling would favor the return of membrane over fluid, leading to the intracellular accumulation of any macromolecule no longer bound to a receptor on the endosomal membrane.[1,19] Adaptor-mediated specificity could clearly be superimposed on this underlying mechanism (see below).

Endosome Subpopulations

From the earliest stages, it was suspected that endosomes did not comprise a simple, homogeneous population but rather a collection of compartments with distinct properties and functions. Early indications came from morphological analyses: endosomes labeled after a brief (1-2 min) uptake of endocytic tracers were generally found in the cell periphery, but then appeared in more centrally located, often multivesicular structures prior to arrival in hydrolase-rich lysosomes. Further, certain recycling receptors, notably Tfn receptor, were in many cell types found to be concentrated in the perinuclear area, close to the Golgi complex and microtubule organizing center.[20,21] Unlike the peripheral compartment, these structures labeled at relatively long times after endocytosis.

The advent of molecular markers for distinct organelle subsets—lgp/lamp, mannose-6-phosphate receptors (MPR), Rab proteins, SNAREs—helped clarify the functions and interrelationships of these various endosomal populations. From cell fractionation experiments, several important concepts were clarified, while others created. We found, for example, that it was possible to physically separate two distinct endosome populations that labeled early (1-2 min) or late (10-15 min) after the uptake of fluid-phase markers. Designated early and late endosomes, both populations were found to be acidic (although late endosomes were more acidic) but otherwise distinct in composition: early endosomes contained recycling receptors and little in the way of lysosomal markers (lgp/lamp, MPR) while late endosomes had the opposite phenotype.[18] Detailed immunofluorescence and immuno-EM experiments soon established definitions for each compartment: early endosomes (MPR and lgp/lamp-negative, recycling receptor-positive), late endosomes (MPR and lgp/lamp-positive), and lysosomes (MPR-negative, lgp/lamp-positive).[22,23] As Rab proteins began to be described, these too were incorporated into the functional definition of endosome compartments, although they were often found in more than one species of endosome (e.g., Rab4/5, early endosomes; Rab7/9, late endosomes).[24-26]

Of special interest was Rab11 (recently joined by Rab8 and others) as marking that subpopulation of Tfn receptor-containing endosomes near the MTOC.[27,28] With this observation, these "recycling endosomes" took on the significance as a third endosome compartment, as opposed to a simple population of transport vesicles. Recycling endosomes always appeared distinct from early endosomes in that they generally did not contain detectable amounts of dissociated ligands or fluid phase markers in transit to lysosomes.[21] Kinetic analysis revealed another difference: although recycling endosomes contained recycling receptors, the recycling endosome pool took far longer to return to the plasma membrane (20-30 min) than the pool that reached only early endosomes (3-4 min).[20,29]

Recycling endosomes became viewed as containing an intracellular pool of recycling membrane components that can be pressed into service when needed, such as possibly providing extra membrane for particle uptake during phagocytosis[30] or possibly during directed cell migration. These intriguing structures may also have essential roles in signal transduction, regulation, and secretion. Their proximity to the Golgi complex continues to cause great confusion as to whether a given marker or event is localized to the Golgi or to endosomes. Indeed, the distinction between recycling endosomes and terminal Golgi elements such as the trans-Golgi network (TGN) may be more semantic than instructive, as will be discussed below.

Endosome Maturation

One of the popular controversies during the first 15 years of endosome research was the issue of endosome biogenesis. Where do they come from? Although the problem has not been entirely solved, the topic appears to have achieved a quiet equilibrium. Helenius et al[1] mused that there were two possible scenarios (Fig. 2). In the maturation model, endosomes might gradually be transformed into lysosomes by virtue of reciprocal fusion events with other organelles and transport vesicles, coupled with the selective recycling of receptors. In the vesicle shuttle model, endosomes (or their subpopulations) were viewed as stable organelles whose contents passed between them via distinct transport vesicles, much as has been understood for transport across the Golgi stack. The model was included not because we felt there was any particular evidence for it in the case of endosomes, but rather due to the strong influence of prevailing views of the Golgi complex. As an aside, it is amusing to recall the impatience expressed by colleagues studying the Golgi who wondered why the endosome community was unable to distinguish between these two models. How the tables have turned!

Considering the considerable capacity for endosomes to move in the cytoplasm, as evidenced from the earliest time lapse video microscopy, it always seemed likely that endosomes were highly dynamic structures that "matured" one into the next. Indeed, direct fusions of endosomes with lysosomes could be visualized.[8] A variety of other more direct considerations now also support a view more in accord with gradual maturation, at least on the lysosomal pathway. For example, the accumulation of multivesicular inclusions pathognomonic of late

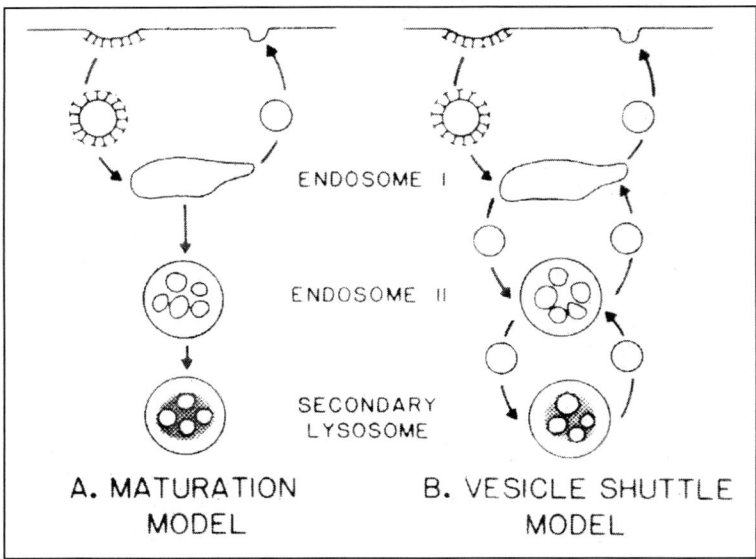

Figure 2. Two views of endosome biogenesis: maturation vs. vesicle shuttle.Reprinted from reference 1, ©1983, with permission from Elsevier.

endosomes and lysosomes begins with the functional assembly of ESCRT complexes at the surface of early endosomes.[31] Combined with the concerted addition of new lysosomal components,[32] the gradual accumulation of multivesicular inclusions emphasizes a remodeling process that converts early endosomes into late endosomes. That the entire endosomal system could turn over rapidly was demonstrated by the nearly complete (and reversible) loss of endosomes in cells where clathrin-mediated endocytosis was arrested by expression of a dynamin GTPase mutant.[33] Thus, endosomes cannot be considered as preexisting stable structures if their presence depends on continuous membrane input by endocytosis. Combined with the prolific amounts of membrane known to move through the system each hour,[8] it is almost a semantic impossibility to consider the endosomal apparatus as being anything other than subject to dynamic remodeling, i.e., maturation.

At the same time, it must also be true that at least some specific transport events take place. The removal of receptors for delivery to the plasma membrane or recycling endosomes must reflect the formation of transport vehicles (vesicles or tubules). Similarly, the return of MPR to the Golgi complex where it must reside in order to capture newly synthesized lysosomal enzymes must also involve a selective recapture pathway.[26] Much the same can be said for the selective return of TGN proteins (TGN38, furin) from endosomes to the Golgi complex.[34,35]

From such considerations, our current view has emerged of endosomes as highly dynamic structures that are closely interrelated and that at least to some degree "mature" from early endosomes to late endosomes to lysosomes. Certainly, some selective sorting or vesicle formation events are also likely to occur, but these do not seem to be stable structures that communicate with each other via a system of small, transport vesicles.

Acidification

It is obvious that a key feature of endosomes is their acidic pH. In general, the farther one proceeds towards lysosomes, the lower the pH. Thus, depending on the pH dissociation profile of a given receptor-ligand complex or fusion threshold of a given enveloped virus, dissociation or infection will occur in different endosomal compartments. Early endosomes are generally given as having a pH of 6-6.8, late endosomes 5-6, and lysosomes 4.5-5, although these numbers probably vary considerably in different cell types.[36]

As mentioned above, it was established early on that endosomes (and lysosomes) contained a proton ATPase. Based on the inhibitor profile of the pump,[12] it was predicted that it would be a member of a unique class of ATPase dedicated to the acidification of both endocytic and secretory organelles. Indeed, subsequent work revealed that the class of "vacuolar ATPase" (V-ATPase) was unique, but was nevertheless closely related to the large F1-F0-like proton ATPase of acidophilic bacteria.[37] Consisting of the same general organization as all such pumps (a soluble multisubunit V1 sector containing the ATPase portion, a membrane-associated multisubunit V0 sector, containing the proton pore), different cells and possibly even different organelles contain different combinatorial forms of the ATPase. The functional significance of such heterogeneity is unclear.

Why early endosomes have a less acidic pH than late endosomes is still not known for certain, but several factors no doubt contribute. The number of pumps may be a factor, with their highest enrichment potentially being in lysosomes. Ion permeabilities of organelle membranes is also different, and is quite likely to contribute to equilibrium pH (Fig. 3). The V-ATPase is electrogenic, meaning that inward pumping of protons is accompanied by the accumulation of an interior-positive membrane potential which, importantly, impedes proton transport probably by directly inhibiting the pump. Endosomal membranes are less leaky to other anions (e.g., Cl^-) and alkalai cations (K^+, Na^+) whose movement across membranes dissipates the forming electrical potential, thus limiting the chemical gradient of protons that the V-ATPase can achieve.[38] The endosome membrane is also quite leaky to protons, making acidification dependent on a dynamic flux of protons per unit time, rather than on a specific number of protons translocated per endosome. Other electrogenic pumps (e.g., Na-K-ATPase, not shown in the diagram) may further limit the pH attained in early endosomes by contributing to the interior-positive potential; the Na-K-ATPase translocates 3 Na^+ in for every 2 K^+ out.[39] Pump assembly (i.e., formation of functional V1-V0 complexes) or subunit composition may also contribute to pH heterogeneity.[40]

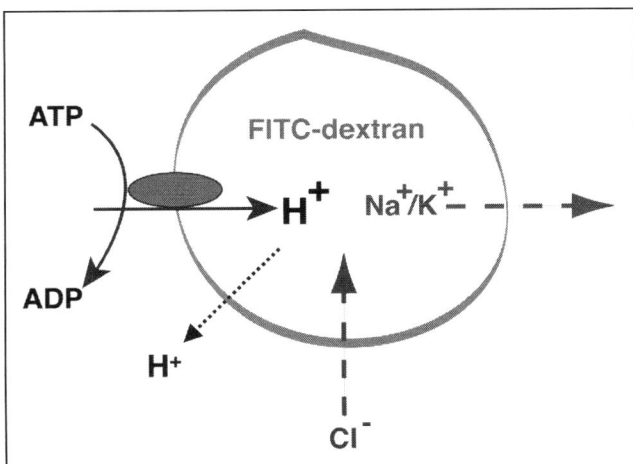

Figure 3. The mechanism of ATP-dependent endosome acidification. Protons are transported into the endosome lumen by the V-ATPase, with their equilibrium concentration determined by the electrochemical gradient reflecting the influx of anions, efflux of alkalai cations, and rate of proton leak back to the cytosol.

Despite the importance of acidification to membrane traffic on the endocytic pathway, the V-ATPase has remained under-studied, perhaps due to its complexity. Given recent results suggesting that its subunits may be directly or indirectly involved in vesicle fusion,[41] and that its assembly can be regulated by certain signaling pathways,[40] further attention would appear warranted.

Sorting Stations at the Crossroads of Membrane Traffic

A primary function of endosomes is molecular sorting, with different endosome compartments playing different roles. Not only are a variety of endocytic sorting events hosted by endosomes, but there are also clear examples where the biosynthetic pathway makes use of endosomal sorting capacity.

Sorting Membrane and Content

The first and perhaps most important sorting function during endocytosis is the separation of membrane from contents, an activity that appears to be the purview of early endosomes. Without this function, cells would not be able to retain or concentrate internalized macromolecules. Early endosomes host both the dissociation of ligands from most rapidly recycling receptors (e.g., LDL receptor) and the physical sorting of the dissociated ligands and other contents in the endosome lumen from membrane and receptors intended for recycling. This is why early endosomes are occasionally referred to as "sorting endosomes" (a term we avoid since all endosomes sort in one way or another). As evidence for sorting of membrane and contents in early endosomes, one need only look at recycling endosomes: it is rare for them to contain fluid-dissolved macromolecules.[21,36] Some fluid does appear to recycle from early endosomes to the extracellular milieu, but this occurs rapidly, and quite possibly reflects the rapid route of recycling that avoids transit through recycling endosomes. Indeed, quantitative measurements have indicated that only 25% of each cohort of internalized receptors reaches recycling endosomes (in MDCK or CHO cells).[29] In any event, it has long been clear, based on quantitative EM measurements, that the concentration of internalized solutes increases as one moves towards later compartments.[8] Thus, endosomes must also sort fluid from solutes by allowing water to egress across the endosomal membrane.

Down Regulation

Since the early work of Cohen,[42] it was apparent that inclusion of receptors in forming MVB's was associated with receptor down regulation.[43] By being removed from the endosome's limiting membrane, the receptors are effectively converted into endosomal content and as such can be taken to lysosomes. Great strides have been made in understanding the biochemistry and genetics of these events both in yeast and animal cells.[44] As will be considered elsewhere in this volume, it is now understood that receptors are typically marked for MVB inclusion by mono-ubiquitination (often following receptor activation following ligand binding or cross-linking). This modification is recognized by proteins such as Hrs (in animal cells) that then trigger recruitment of the ESCRT complex. These complexes initiate the invagination of the endosome's limiting membrane, sequestering the ubiquitin-marked receptors. Since MVBs are classified morphologically as late endosomes, it was presumed that late endosomes were the primary site for this sorting event. While that may be true, as mentioned earlier ESCRT complex proteins, as well as proteins involved in recognizing the mono-ubiquitination signals associated with receptor down regulation first bind to early endosomes.[31] Thus, early endosomes may be progressively converted into late endosomes by concerted MVB formation.

Lysosomal Biogenesis

It has long been known that MPR's carry newly synthesized lysosomal enzymes from the TGN to early and/or late endosomes. Upon delivery, the acidic pH facilitates dissociation of the receptor-ligand complex resulting in the delivery of the discharged enzymes to lysosomes, along with any internalized content.[32] By making use of a selective targeting event initiated at the TGN, the cell basically makes use of endosomal sorting to serve a specific function on the secretory pathway.

MHC Class II Molecules and Antigen Presentation

Another example of an intersection between the biosynthetic and endocytic pathways can be found in antigen presenting cells of the immune system.[45] Here, newly synthesized MHC class II molecules, in association with their targeting chaperone "invariant chain", are targeted from the TGN to endosomes (early or late?), where the complexes can exhibit one of two fates. If the invariant chain is cleaved by endosomal-lysosomal proteases (typically cathepsin S), the released αβ dimers can proceed (via recycling) to the plasma membrane. If not, then the entire complex is transferred to lysosomes, likely following inclusion on forming MVB's.[46] This pathway is an essential aspect of the immune response. It is in endocytic compartments that newly synthesized MHC class II molecules encounter exogenous antigens internalized by endocytosis, and thus acquire the 10-15-mer peptides that are required for presentation to T lymphocytes. The ability to mount immunity to foreign antigens and tolerance to self antigens is therefore intimately dependent on the sorting functions of endosomes, a point illustrated dramatically in dendritic cells, the immune systems most important antigen presenting cell.[47]

Endosomal Sorting in Polarized Cells

Endosomal sorting and recycling in polarized epithelial cell presents an additional challenge: cells must maintain the ability to selectively return incoming receptors to the basolateral or apical surfaces. Interestingly, endosomes accomplish this task by recognizing the same cytoplasmic and luminal sorting signals on basolateral and apical proteins as are recognized in the secretory pathway.[48,49] Thus, endosomes must be capable of signal-dependent sorting, and not just the bulk separation of recycling receptors from endosomal contents or lysosomally-directed components.

Given the equivalence of signals used, it is possible that polarized sorting in endosomes uses the same sets of adaptors as polarized sorting on the secretory pathway. In the case of tyrosine-based basolateral targeting signals, one complex may be the AP-1B clathrin adaptor.[50] At present, the best evidence supports the possibility that signal-dependent sorting occurs uniquely in recycling endosomes (Fig. 4). Although not yet conclusive, kinetic analysis suggests that signals control sorting only in the slowly recycling pool of receptors such as transferrin receptor, apparently at the level of recycling endosomes in MDCK cells.[29] Moreover, during transcytosis, a common pool of recycling endosomes may receive input from both the apical and basolateral domains,[29] although there is some immunofluorescence evidence suggesting an additional, specialized endosomal compartment that fulfills this role.[51]

More recently, we have found that many of the components which in polarized cells are essential for basolateral transport on the secretory pathway actually localize to recycling endosomes in MDCK cells.[28,52] Indeed, both immuno-EM and functional evidence indicates that recycling endosomes serve as an intermediate on the biosynthetic pathway to the surface.[53] Thus, there may be a common site for all polarized sorting in epithelial cells, and that site would represent yet another remarkable convergence of the endocytic and secretory pathways.

Endosomes in Specialized Cells

The example of polarized epithelial cell raises the issue of whether specialized cell types exhibit endosome specializations that serve cell type-specific functions. Although unresolved, there is considerable evidence that this may be the case. In adipocytes, for example, vesicles containing the Glut4 glucose transporter may form by a specific budding event from an endosomal intermediate.[54,55] Much the same idea has been proposed in the case of synaptic vesicle formation in neuroendocrine cells.[56] In such cases, one can imagine that all of the components required for the formation of specialized vesicles are delivered by endocytosis to endosomes, at which site they are sorted and sequestered into populations of recycling vesicles whose recruitment to the plasma membrane can be carefully regulated.

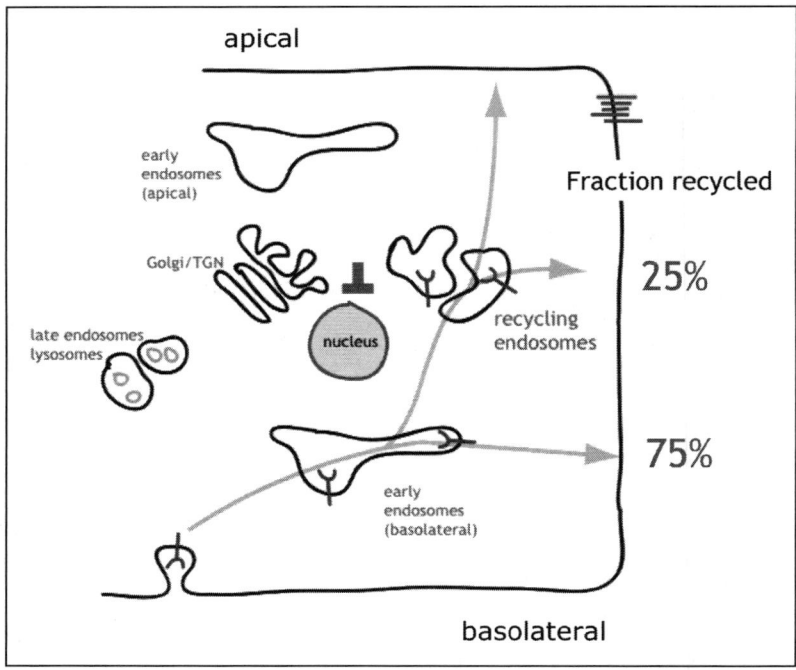

Figure 4. Proposed scheme of sorting by recycling endosomes in polarized cells. As in non-polarized cells, only ~25% of each internalized cohort of receptors actually reaches the perinuclear recycling endosomes. Here, they are interrogated for the presence of basolateral targeting signals by adaptor complexes such as AP-1B. If a productive interaction occurs, these receptors are efficiently recycled back to the basolateral surface. If not, they are transported from recycling endosomes to the apical surface, thereby exhibiting "transcytosis". Recycling traffic from early endosomes in the basolateral cytoplasm is thought to be largely to the basolateral surface, but is signal-independent: transport occurs regardless of whether a specific basolateral targeting signal is present. See reference 29.

In antigen presenting cells, identifying the actual site(s) at which antigen is degraded and loaded onto MHC class II molecules has long been a problem of interest. Initial work characterized a late endosome-like compartment (MIIC) which was thought to represent an organelle specialized for peptide loading, but are now thought simply to reflect "conventional" multivesicular late endosomes that happen to be loaded with MHC class II and associated molecules.[57-59] Whether, in B cells, these structures represent the major loading site, however, remains unclear.

With the demonstration that dendritic cells play by far the most important role in initiating antigen-specific immune responses, attention has recently been shifted to the endocytic system in this cell type. Although unremarkable at first glance, dendritic cells possess the ability to completely reorganize their endosomes and lysosomes in response to stimuli that activate their antigen processing and presentation abilities.[47] Targeting of MHC class II is altered, lysosomal acidification is activated due to V-ATPase assembly, and peptide-MHC class II complexes are induced to form. Strikingly, the late endosomal/lysosomal structures that contain the bulk of MHC class II molecules in resting "immature" cells tubulates, and transfers its MHC class II to the plasma membrane.[59-62] This rearrangement involves the loss of MHC class II-positive internal vesicles from these MVB-like structures, raising the possibility that they somehow have fused with the endosome's limiting membrane (or were degraded upon activation of dendritic cell protease activity that also occurs upon maturation).[40] If the internal vesicles did indeed fuse, this would be a totally unexpected fate for a class of membranes until now thought to be marked for degradation.

Conclusions: Endosomes in Cell Biology

In this brief review, we have tried to place our burgeoning understanding of endosomes in the context of their discovery and initial characterization, not so very long ago. We have explored no more than a few of the most striking functional attributes and molecular insights that have emerged. For example, one key function pertains to the role of endosomes in signal transduction. The surface expression of signaling receptors can be rapidly modulated by regulating traffic to and from the recycling endosome pool. These same structures may also serve as platforms for generating signals. Similarly, in the TGFβ pathway, critical adaptor molecules are found only in early endosomes, meaning that signaling cannot occur until the receptor is internalized.[63,64]

The efforts of a great many groups spanning two decades has established endosomes as key and important organelles in a wide range of cell types and pathways. This work has contributed to our understanding of fundamental cell biology in other ways, however. Endosomes have provided a remarkable convenient platform on which to develop concepts concerning the function of Rab proteins, effectors, SNAREs and tethers in processes ranging from the formation of membrane microdomains to membrane fusion.[65] Arguably, more has been learned in this respect from the study of endosomes than from any other organelle.

Endosomes may have come of age quickly, but they have done it well.

Acknowledgments

I am completely indebted to the many wonderful laboratory members with whom I have had the privilege of working in my years at Rockefeller and Yale Medical School. It is gratifying that so many of them have gone on to be major contributors to the field. I am also indebted to the wonderful group of colleagues with whom I have interacted these many years; they are more responsible for my ideas than I am myself. In particular, I must single out my partner, palindromic brother, and companion Ari Helenius for his support, encouragement, and unabiding friendship for 25 years. My current lab partner, Graham Warren, also deserves immense credit and deep thanks for keeping me intellectually alive and sharp since Ari's departure for Zürich several years ago. Special thanks, also, to the late Zanvil Cohn who gave me the freedom and encouragment to be ambitious, and to my close friend Ralph Steinman who was there with me at the very beginning, and with whom I will remain until the end. Support for our efforts have been generously provided by a variety of sources, most notably the NIH and the Ludwig Institute for Cancer Research, a unique organization dedicated to innovation in the service of cancer research.

References

1. Helenius A, Mellman I, Wall D et al. Endosomes. Trends in Biochem Sci 1983; 8:245-250.
2. Steinman RM, Brodie SE, Cohn ZA. Membrane flow during pinocytosis: A stereologic analysis. J Cell Biol 1976; 68(3):665-687.
3. Steinman RM, Silver JM, Cohn ZA. Pinocytosis in fibroblasts: Quantitative studies in vitro. J Cell Biol 1974; 63(3):949-969.
4. Anderson RG, Goldstein JL, Brown MS. Localization of low density lipoprotein receptors on plasma membrane of normal human fibroblasts and their absence in cells from a familial hypercholesterolemia homozygote. Proc Natl Acad Sci USA 1976; 73(7):2434-2438.
5. Anderson RG, Brown MS, Goldstein JL. Role of the coated endocytic vesicle in the uptake of receptor-bound low density lipoprotein in human fibroblasts. Cell 1977; 10(3):351-364.
6. Goldstein JL, Anderson RG, Brown MS. Coated pits, coated vesicles, and receptor-mediated endocytosis. Nature 1979; 279(5715):679-685.
7. Mellman IS, Steinman RM, Unkeless JC et al. Selective iodination and polypeptide composition of pinocytic vesicles. J Cell Biol 1980; 86(3):712-722.
8. Steinman RM, Mellman IS, Muller WA et al. Endocytosis and the recycling of plasma membrane. J Cell Biol 1983; 96(1):1-27.

9. Marsh M, Bolzau E, Helenius A. Penetration of Semliki Forest virus from acidic prelysosomal vacuoles. Cell 1983; 32(3):931-940.

10. van Renswoude J, Bridges KR, Harford JB et al. Receptor-mediated endocytosis of transferrin and the uptake of fe in K562 cells: Identification of a nonlysosomal acidic compartment. Proc Natl Acad Sci USA 1982; 79(20):6186-6190.

11. Tycko B, Maxfield FR. Rapid acidification of endocytic vesicles containing alpha 2-macroglobulin. Cell 1982; 28(3):643-651.

12. Galloway CJ, Dean GE, Marsh M et al. Acidification of macrophage and fibroblast endocytic vesicles in vitro. Proc Natl Acad Sci USA 1983; 80(11):3334-3338.

13. Robbins AR, Oliver C, Bateman JL et al. A single mutation in Chinese hamster ovary cells impairs both Golgi and endosomal functions. J Cell Biol 1984; 99(4 Pt 1):1296-1308.

14. Geuze HJ, Slot JW, Strous GJ et al. Intracellular site of asialoglycoprotein receptor-ligand uncoupling: Double-label immunoelectron microscopy during receptor-mediated endocytosis. Cell 1983; 32(1):277-287.

15. Mellman I, Plutner H. Internalization and degradation of macrophage Fc receptors bound to polyvalent immune complexes. J Cell Biol 1984; 98(4):1170-1177.

16. Mellman I, Plutner H, Ukkonen P. Internalization and rapid recycling of macrophage Fc receptors tagged with monovalent antireceptor antibody: Possible role of a prelysosomal compartment. J Cell Biol 1984; 98(4):1163-1169.

17. Marsh M, Schmid S, Kern H et al. Rapid analytical and preparative isolation of functional endosomes by free flow electrophoresis. J Cell Biol 1987; 104(4):875-886.

18. Schmid SL, Fuchs R, Male P et al. Two distinct subpopulations of endosomes involved in membrane recycling and transport to lysosomes. Cell 1988; 52(1):73-83.

19. Marsh M, Griffiths G, Dean GE et al. Three-dimensional structure of endosomes in BHK-21 cells. Proc Natl Acad Sci USA 1986; 83(9):2899-2903.

20. Hopkins CR. Intracellular routing of transferrin and transferrin receptors in epidermoid carcinoma A431 cells. Cell 1983; 35(1):321-330.

21. Yamashiro DJ, Tycko B, Fluss SR et al. Segregation of transferrin to a mildly acidic (pH 6.5) para-Golgi compartment in the recycling pathway. Cell 1984; 37(3):789-800.

22. Griffiths G, Hoflack B, Simons K et al. The mannose 6-phosphate receptor and the biogenesis of lysosomes. Cell 1988; 52(3):329-341.

23. Geuze HJ, Stoorvogel W, Strous GJ et al. Sorting of mannose 6-phosphate receptors and lysosomal membrane proteins in endocytic vesicles. J Cell Biol 1988; 107(6 Pt 2):2491-2501.

24. Chavrier P, Parton RG, Hauri HP et al. Localization of low molecular weight GTP binding proteins to exocytic and endocytic compartments. Cell 1990; 62(2):317-329.

25. Van Der Sluijs P, Hull M, Zahraoui A et al. The small GTP-binding protein rab4 is associated with early endosomes. Proc Natl Acad Sci USA 1991; 88(14):6313-6317.

26. Lombardi D, Soldati T, Riederer MA et al. Rab9 functions in transport between late endosomes and the trans Golgi network. EMBO J 1993; 12(2):677-682.

27. Ullrich O, Reinsch S, Urbe S et al. Rab11 regulates recycling through the pericentriolar recycling endosome. J Cell Biol 1996; 135(4):913-924.

28. Ang AL, Folsch H, Koivisto UM et al. The Rab8 GTPase selectively regulates AP-1B-dependent basolateral transport in polarized Madin-Darby canine kidney cells. J Cell Biol 2003; 163(2):339-350.

29. Sheff DR, Daro EA, Hull M et al. The receptor recycling pathway contains two distinct populations of early endosomes with different sorting functions. J Cell Biol 1999; 145(1):123-139.

30. Bajno L, Peng XR, Schreiber AD et al. Focal exocytosis of VAMP3-containing vesicles at sites of phagosome formation. J Cell Biol 2000; 149(3):697-706.

31. Raiborg C, Bache KG, Gillooly DJ et al. Hrs sorts ubiquitinated proteins into clathrin-coated microdomains of early endosomes. Nat Cell Biol 2002; 4(5):394-398.

32. Kornfeld S, Mellman I. The biogenesis of lysosomes. Annu Rev Cell Biol 1989; 5:483-525.

33. Damke H, Baba T, Warnock DE et al. Induction of mutant dynamin specifically blocks endocytic coated vesicle formation. J Cell Biol 1994; 127(4):915-934.

34. Ghosh RN, Mallet WG, Soe TT et al. An endocytosed TGN38 chimeric protein is delivered to the TGN after trafficking through the endocytic recycling compartment in CHO cells. J Cell Biol 1998; 142(4):923-936.

35. Molloy SS, Thomas L, VanSlyke JK et al. Intracellular trafficking and activation of the furin proprotein convertase: Localization to the TGN and recycling from the cell surface. EMBO J 1994; 13(1):18-33.

36. Mellman I. Endocytosis and molecular sorting. Annu Rev Cell Dev Biol 1996; 12:575-625.

37. Nishi T, Forgac M. The vacuolar (H+)-ATPases—nature's most versatile proton pumps. Nat Rev Mol Cell Biol 2002; 3(2):94-103.

38. Fuchs R, Male P, Mellman I. Acidification and ion permeabilities of highly purified rat liver endosomes. J Biol Chem 1989; 264(4):2212-2220.
39. Fuchs R, Schmid S, Mellman I. A possible role for Na+,K+-ATPase in regulating ATP-dependent endosome acidification. Proc Natl Acad Sci USA 1989; 86(2):539-543.
40. Trombetta ES, Ebersold M, Garrett W et al. Activation of lysosomal function during dendritic cell maturation. Science 2003; 299(5611):1400-1403.
41. Hiesinger PR, Fayyazuddin A, Mehta SQ et al. The v-ATPase V0 subunit a1 is required for a late step in synaptic vesicle exocytosis in Drosophila. Cell 2005; 121(4):607-620.
42. Haigler HT, McKanna JA, Cohen S. Direct visualization of the binding and internalization of a ferritin conjugate of epidermal growth factor in human carcinoma cells A-431. J Cell Biol 1979; 81(2):382-395.
43. Felder S, Miller K, Moehren G et al. Kinase activity controls the sorting of the epidermal growth factor receptor within the multivesicular body. Cell 1990; 61(4):623-634.
44. Katzmann DJ, Odorizzi G, Emr SD. Receptor downregulation and multivesicular-body sorting. Nat Rev Mol Cell Biol 2002; 3(12):893-905.
45. Neefjes JJ, Stollorz V, Peters PJ et al. The biosynthetic pathway of MHC class II but not class I molecules intersects the endocytic route. Cell 1990; 61(1):171-183.
46. Kleijmeer M, Ramm G, Schuurhuis D et al. Reorganization of multivesicular bodies regulates MHC class II antigen presentation by dendritic cells. J Cell Biol 2001; 155(1):53-63.
47. Trombetta ES, Mellman I. Cell biology of antigen processing in vitro and in vivo. Annu Rev Immunol 2005; 23:975-1028.
48. Matter K, Whitney JA, Yamamoto EM et al. Common signals control low density lipoprotein receptor sorting in endosomes and the Golgi complex of MDCK cells. Cell 1993; 74(6):1053-1064.
49. Aroeti B, Mostov KE. Polarized sorting of the polymeric immunoglobulin receptor in the exocytotic and endocytotic pathways is controlled by the same amino acids. EMBO J 1994; 13(10):2297-2304.
50. Folsch H. The building blocks for basolateral vesicles in polarized epithelial cells. Trends Cell Biol 2005; 15(4):222-228.
51. Wang E, Brown PS, Aroeti B et al. Apical and basolateral endocytic pathways of MDCK cells meet in acidic common endosomes distinct from a nearly-neutral apical recycling endosome. Traffic 2000; 1(6):480-493.
52. Folsch H, Pypaert M, Maday S et al. The AP-1A and AP-1B clathrin adaptor complexes define biochemically and functionally distinct membrane domains. J Cell Biol 2003; 163(2):351-362.
53. Ang AL, Taguchi T, Francis S et al. Recycling endosomes can serve as intermediates during transport from the Golgi to the plasma membrane of MDCK cells. J Cell Biol 2004; 167(3):531-543.
54. Bryant NJ, Govers R, James DE. Regulated transport of the glucose transporter GLUT4. Nat Rev Mol Cell Biol 2002; 3(4):267-277.
55. Wei ML, Bonzelius F, Scully RM et al. GLUT4 and transferrin receptor are differentially sorted along the endocytic pathway in CHO cells. J Cell Biol 1998; 140(3):565-575.
56. de Wit H, Lichtenstein Y, Geuze HJ et al. Synaptic vesicles form by budding from tubular extensions of sorting endosomes in PC12 cells. Mol Biol Cell 1999; 10(12):4163-4176.
57. Pierre P, Denzin LK, Hammond C et al. HLA-DM is localized to conventional and unconventional MHC class II-containing endocytic compartments. Immunity 1996; 4(3):229-239.
58. Pierre P, Mellman I. Exploring the mechanisms of antigen processing by cell fractionation. Curr Opin Immunol 1998; 10(2):145-153.
59. Kleijmeer MJ, Morkowski S, Griffith JM et al. Major histocompatibility complex class II compartments in human and mouse B lymphoblasts represent conventional endocytic compartments. J Cell Biol 1997; 139(3):639-649.
60. Boes M, Cerny J, Massol R et al. T-cell engagement of dendritic cells rapidly rearranges MHC class II transport. Nature 2002; 418(6901):983-988.
61. Chow A, Toomre D, Garrett W et al. Dendritic cell maturation triggers retrograde MHC class II transport from lysosomes to the plasma membrane. Nature 2002; 418(6901):988-994.
62. Chow AY, Mellman I. Old lysosomes, new tricks: MHC II dynamics in DCs. Trends Immunol 2005; 26(2):72-78.
63. Panopoulou E, Gillooly DJ, Wrana JL et al. Early endosomal regulation of Smad-dependent signaling in endothelial cells. J Biol Chem 2002; 277(20):18046-18052.
64. Hayes S, Chawla A, Corvera S. TGF beta receptor internalization into EEA1-enriched early endosomes: Role in signaling to Smad2. J Cell Biol 2002; 158(7):1239-1249.
65. Zerial M, McBride H. Rab proteins as membrane organizers. Nat Rev Mol Cell Biol 2001; 2(2):107-117.

CHAPTER 2

Lipid Membrane Domains in Endosomes

Julien Chevallier and Jean Gruenberg*

Abstract

It has long been appreciated that the membranes of endosomes contain different regions or domains visible by electron microscopy, including, for example, the intraluminal and limiting membranes of multivesicular compartments. Evidence also shows that endosomes contain different lipid territories, and that such territories overlap with morphologically visible domains. Here, we will discuss recent advances in our understanding of the role of these specialized membrane domains and protein-lipid assemblies in the endocytic pathway leading to lysosomes.

Introduction

Eukaryotic cells need to be in constant communication with their environment in order to perform most of their functions, such as the transmission of neuronal, metabolic, and proliferative signals, the uptake of nutrients, or to protect the organism from microbial invasion, to name only a few. During endocytosis, cell surface receptors and their ligands, as well as particles or solutes present in the extracellular space, are taken up by vesicles that form at the plasma membrane, sorted to early endosomes, and then targeted to various intracellular destinations (Fig. 1). As a consequence, the lumen of endosomes—and of all organelles of the vacuolar apparatus—is topologically equivalent with the extracellular space. Lysosomes are a common final destination for endocytosed macromolecules, where digestive enzymes degrade them. The resulting metabolites are then released into the cytoplasm where they can be recycled by incorporation into newly synthesised macromolecules.

Endosomes (like biosynthetic organelles) exhibit a wide variety of shapes and structures, which can be easily visualised by classical electron microscopy. They range from the clusters of thin, long tubules of recycling endosomes to late endosomes that contain onion-like sheets of internal membranes, tubules or vesicles (multivesicular or multilamellar endosomes). While little is known about the molecular mechanisms controlling organelle shape and biogenesis, or the functional significance of such diversity, evidence shows that organelles in the endocytic pathway are composed of a mosaic of structural and functional regions.[1,2] These regions consist, at least in part, of specialized protein–lipid domains within the plane of the membrane or of protein complexes associated to specific membrane lipids. Indeed, many cytosolic proteins interact with membranes by binding not only to proteins but also to lipids, often through multiple protein-lipid and protein-protein interactions.[3-5] Such interactions are not easily studied, however, and it should be emphasized that physiologically-relevant parameters, e.g., kinetic constants, are not always known. In any case, the dynamic interplay between such specialized

*Corresponding Author: Jean Gruenberg—Department of Biochemistry, Sciences II, 30 quai E. Ansermet, 1211 Geneva 4, Switzerland. Email: jean.gruenberg@biochem.unige.ch

Endosomes, edited by Ivan Dikic. ©2006 Landes Bioscience and Springer Science+Business Media.

protein–lipid domains may provide a driving force responsible both for the specific organization of each compartment and for the movement of cargo molecules.

Lipids provide the physical support of organelle membranes, acting as a barrier for water-soluble molecules and as a solvent for the hydrophobic domains of membrane proteins. By contributing to the intrinsic properties of membranes, such as thickness, asymmetry, and curvature, lipids can potentially regulate protein movement and distribution.[6,7] Evidence is accumulating that some short- and long-lived lipids have a restricted distribution in the plane of the bilayer, thereby forming transient or more stable microdomains.[8] In particular, cholesterol and sphingolipids were proposed to form a separate liquid-ordered phase in the liquid-disordered matrix of the lipid bilayer (lipid rafts), thereby functioning as platforms that can incorporate distinct classes of proteins, and thus regulate numerous cellular processes, including signalling, sorting and infection.[9-11] Here, we will discuss the organization of endosomes into different membrane domains, and, in particular, evidence supporting the notion that, in animal cells, endosomes along the degradation pathway leading to lysosomes contain more than one type of membrane domains with different lipid compositions and functions.

Mophologically-Visible Domains

As mentionned above, it has long been appreciated that endosomes contain specialized membrane regions or domains that are visible by electron microscopy. While thin tubules form the elements of recycling endosomes, all endosomes along the degradation pathway leading to lysosomes can accumulate internal membranes in their lumen, thus appearing multivesicular or multilamellar. At early stage of the latter degradation pathway in animal cells, multivesicular endosomes form regularly-shaped and large vesicles (diameter $\approx 0.4 - 0.5$ μm) with densely packed intraluminal membranes that appear like small vesicles or tubules (diameter: 50-80 nm).[12] Such multivesicular endosomes form on early endosomal membranes and mediate transport to late endosomes, and have thus been referred to as endosomal carrier vesicles (ECVs) or multivesicular bodies (MVBs) according to their function or appearance, respectively.

By contrast with ECV/MVBs, the large vesicular elements of late endosomes are often less regularly-shaped, with sizes ranging from ≈ 0.5 to 1.0 μm, and can exhibit a more intraluminal organization, including internal vesicles or tubules (like ECV/MVBs), onion-like sheets (multilamellar endosomes), or a mixture of both. In some cells, late endosomes also accumulate electron-lucent materials, perhaps of lipidic origin that can form elongated, needle-like structures (e.g., BHK cells and AtT20 cells). Clearly, this bird's eye view of these organelles reveals that, beyond all mechanistic debates, endosomes along the degradation pathway contain different morphologically-visible membrane regions or domains.

Lipid Distribution

All lipids do not behave as bulk constituents of the bilayer and are not all stochastically distributed within membranes of endocytic organelles.[4,8,13] In particular, studies with toxins that bind cell surface glycolipids,[14] or with fluorescent lipid analogs[8,15,16] indicate that different lipids or lipid analogs inserted into the plasma membrane may follow different intracellular routes after endocytosis. In addition, evidence is accumulating that endosomal membranes also contain different lipids at successive stage of the degradation pathway.

Over the past few years, phosphoinositides have emerged as key-regulators of membrane traffic by controlling the localisation and/or activity of effector proteins, through the action of kinases and phosphatases that mediate highly localised changes in the level of phosphoinositides, providing a means for the temporal and spatial regulation of effectors.[3,7] In the endocytic pathway, phosphatidylinositol 4,5-bisphosphate, PI(4,5)P$_2$, plays a crucial role during internalization, by recruiting proteins implicated in endocytosis, including the AP-2 adaptor, the GTPase dynamin, and proteins that contain an ENTH (Epsin NH2-Terminal Homology)-like domain, e.g., CALM (clathrin assembly lymphoid myeloid leukemia protein), AP180 and Epsin.[3] In addition to PI(4,5)P$_2$, phosphatidylinositol 3-phosphate, PI(3)P, also regulates endocytic membrane traffic, but presumably at the next step of the pathway, on early endosomes. PI(3)P is generated at least in part on early endosomal membranes via the recruitment of the PI3K hVPS34 by the active GTP-bound Rab5, and thus contributes to the formation of Rab5 effector platforms.[2] PI(3)P plays a major role in endocytic traffic through interactions with the FYVE zinc finger domain that is present in over 10 different proteins, including Rab5 effectors, with a wide range of structures and functions in mammalian cells. In addition, the human genome also contains many ($\approx$ 50) genes that encode proteins with the phosphoinositide-binding Phox homology (PX) domain. Amongst those that have been characterized, many PX-proteins bind PI(3)P, in particular some members of the sorting nexin family.[17,18] Interestingly, labelling of cryo-sections with a tandem-FYVE construct revealed that PI(3)P is abundant in the internal membranes of ECV/MVBs, and, to a much lesser extent, of late endosomes.[19] PI(3)P also serves as a substrate for the PtdIns3P 5-kinase Fab1/PIKfyve that generates PtdIns(3,5)P$_2$. While Fab1/PIKfyve and its product PtdIns(3,5)P$_2$ clearly play a crucial role in protein sorting,[20-24] the precise localization of the lipid in endosomal membrane is not known.

A very similar distribution was observed for cholesterol, when probing cryo-sections with a derivative of the cholesterol-binding Theta-toxin (perfringolysin O),[25] perhaps suggesting that intra-endosomal cholesterol and PI3P are both abundant within the same ECV/MVB internal membranes (see Fig 1, green membranes). But, it is also possible that the two lipids distribute preferentially within different pools of internal vesicles. By contrast, late endosomes accumulate large amounts (>15 Mol%) of the unconventional phospholipids lyso-bisphosphatidic acid (LBPA) or bis-monoacylglycerophosphate (BMP), and this lipid is not detected elsewhere in the cell.[26] Immunogold labelling of cryosections with anti-LBPA antibodies shows that the lipid is abundant in internal membranes,[26] where it does not seem to colocalize with PI3P[19] or cholesterol[25] (see Fig 1, red membranes). Altogether these studies indicate that endosomes along the degradation pathway in mammalian cells may contain at least two types of intraluminal membranes with different lipid compositions, including perhaps some enriched in PI3P-cholesterol and LBPA, respectively (Fig. 1, outline).

It should be emphasized that it has not been possible until now to correlate the different morphologically visible regions of late endosome internal membranes (e.g., multivesicular vs. multilamellar) with differences in biophysical or biochemical properties. However, the sub-organellar fractionation of late endosomes revealed not only that internal membranes could be separated on gradients without detergent from the limiting membrane of the organelle, but also that at least two populations of internal membranes with a different lipid composition can be separated from each other.[27] In addition, late endosomes also contain detergent-resistant membranes (DRMs) enriched in glycosylphosphatidyl inositol (GPI)-anchored proteins and cholesterol, presumably rafts.[28] Recent studies, in fact, suggest that two populations of DRMs with a different protein composition are present in late endosomes, within internal membranes and at the limiting membrane, respectively.[29] Whether intraluminal DRMs correspond to the cholesterol-rich internal membranes visible by electron microscopy[25] is not known. But, these studies indicate that (at least) two populations of internal membranes continue to coexist in late endosomes, further supporting the notions discussed above (see Fig 1).

Functionally Different Membrane Domains

In addition to these differences in morphology and composition, evidence is also accumulating that endosomes in mammalian cells contain more than one population of functionally different membrane domains. Indeed, when some signalling receptors are downregulated in the presence of excess ligand, they are endocytosed and then rapidly appear within the intraluminal vesicles of endosomes, thus providing an efficient means to turn off signalling, by removing the receptor from its interacting signalling partners present in the cytosol.[30,31] Major progress has been made in understanding the molecular mechanisms that control this sorting event, which allows the selective incorporation of receptors destined to be degraded within these intraluminal membrane invaginations. Some downregulated receptors are ubiquitinylated and this modification is responsible for sorting into forming ECV/MVBs through binding to Hrs (hepatocyte growth factor-regulated tyrosine kinase substrate), which also binds PI(3)P via its FYVE domain, and ESCRTs-I, -II and -III (endosomal sorting complexes required for transport).[32-34] Conversely, Hrs and ESCRT complexes are believed to drive membrane invagination itself, since the process is inhibited in yeast and Drosophila mutants with impaired Hrs (VPS28 in yeast) functions, and in mammalian cells treated with Hrs siRNAs.[33,35,36] This mechanism, which is conserved from yeast to mammals, leads to the notion that internal vesicles with their cargo of lipids, downregulated receptors and, presumably, other proteins are transported to lysosomes for degradation. Indeed, intraluminal vesicles accumulate in the vacuole of yeast degradation mutants.[33]

However, intraluminal membranes present in the late endosomes of animal cells also contain proteins and lipids that are not destined for the lysosomes. LBPA, which accumulates within intraluminal membranes, is in fact poorly degradable, perhaps because of its unconventional stereochemistry.[37] Late endosome internal membranes also typically contain members of the tetraspanin family including CD63/Lamp3,[38] which are presumably not destined to be degraded. Moreover, MHC (major histocompatibility complex) class II molecules are predominantly found within internal membranes of late endosomes (MIICs) in dendritic cells. Upon cell activation, these molecules are rapidly transported to the cell surface demonstrating that back-transport from late endosomes internal membranes can occur, at least in these cells, presumably via tubules,[39] that may form at the expense of internal membranes via back-fusion.[12] In addition, the mannose-6-phosphate receptor (MPR), which delivers newly-synthesized lysosomal enzymes to endosomes and lysosomes, cycles between the trans-Golgi network (TGN) and endosomes, with the bulk present in the TGN at steady state in some cell types. While in transit in endosomes, MPR is found within late endosome internal membranes,[40] where it accumulates in cells containing endocytosed antibodies against LBPA.[26] The situation may be different in yeast cells. It is not clear whether yeast cells contain LBPA, and cargoes that recycle from intraluminal vesicles to the limiting membrane of yeast endosomes have not been identified. Even if a related recycling pathway may exist in yeast, it is likely to be of lesser importance than in animal cells. Indeed, although MVEs have been observed in yeast,[41] membrane invaginations and internal vesicles seem to be far more abundant in the endosomes of animal cells, and are readily visible at a steady state. It thus appears that, in addition to the downregulation pathway conserved from yeast to man, animal cells have evolved a more elaborate membrane system in late endosomes for more efficient reutilization and sorting of specialized lipid and protein (see outline, Fig. 1).

Endosomes in animal cells thus seem to contain at least two morphologically, biochemically and functionally different populations of intraluminal vesicles (Fig. 1). It is tempting thus to speculate that the internal vesicles that form on early endosmal membranes via a mechanism involving the short-lived lipid PI3P and the PI3P-binding protein Hrs, as well as ESCRT complexes, correspond to the vesicles that accumulate in ECV/MVBs and contain both cholesterol and PI3P, as well as cargo molecules that need to be degraded, in particular signalling receptors. These vesicles contained in the endosomal lumen, are then presumably transported via late endosomes to the lysosomes for complete degradation. In addition, late endosomes also seem to contain a second population of internal vesicles, which are rich in LBPA, and thus poorly degradable. These membranes contain proteins that are not destined for the lysosomes, but can be returned to the limiting membrane (presumably via back-fusion of the intraluminal vesicles) and then transported to other cellular destinations.

LBPA and Alix/ALP1

While the biophysical and biochemical properties of internal membranes are still poorly understood, progress has been made in understanding some of the properties and functions of late endosome internal membranes rich in LBPA. This lipid is presumably synthesized in situ within the acidic organelles of the endocytic pathway,[42] and, in BHK cells, is predominantly ($\approx$ 90%) present as the 2,2'-dioleoyl isoform[27,43] (Fig. 2). But, the β-position of the glycerol backbone, to which the oleoyl chains are esterified in 2,2'-LBPA, is thermodynamically unstable, and fatty acids can migrate to the α-positions, thus forming 3,3'-LBPA.[43] Such acyl chain migration may well contribute to regulate the function of the lipid in vivo: 2,2-LBPA, but presumably not 3,3'-LBPA, is predicted to be cone shaped,[27,43] and may thus facilitate the formation of membrane invaginations. Indeed, 2,2'-LBPA, but not 3,3'-LBPA, drives the spontaneous formation of multivesicular liposomes, when the liposome lumen is acidified to the pH (≈ 5.5) of late endosomes.[44] Hence, 2,2'-LBPA, the major late endosomal isoform, is endowed with the intrinsic capacity to stimulate internal vesicle formation within acidic liposomes, and thus to generate structures that resemble late endosomes where the lipid is found in vivo.[44] This mechanism is attractive. Invagination occurs towards the endosomal lumen, and is thus unlikely to depend on cytosolic machineries (e.g., coat proteins) that control vesicle formation in the topologically opposite direction.

This invagination process is likely to be regulated by proteins in vivo, and indeed it was found to depend on Alix/ALP1, which, in turn, binds liposomes containing 2,2'-LBPA, but not 3,3'-LBPA.[44] Moreover, Alix downregulation with siRNAs affect both late endosome membrane organization and the cellular LBPA content.[44] Although the precise function of Alix remains to be unravelled, other studies already provide some insights into its biological role.[5] Alix, which was identified as a partner of ALG-2 involved in apoptosis, interacts with proteins that play a role in signalling and endocytosis,[5] and is the mammalian homologue of yeast Bro1p/Vps31p, which regulates MVB formation in concert with ESCRT proteins. Consistently, Alix together with ESCRT proteins play a role in HIV budding at the plasma membrane, presumably reflecting the capacity of the virus to hijack proteins that normally drive the topologically equivalent process of invagination within endosomes.[45,46] Whether Bro1p/Vps31p functions are LBPA-dependent in yeast is not known, since LBPA was not detected in yeast. However, it is possible that Alix acquired the capacity to interact with LBPA later in evolution, since Alix and Bro1p are relatively distantly related (≈ 24% identity by the Jotun Hein method, and 17% by the clustal method).

Intraluminal Traffic

Several lines of evidence indicate that LBPA and Alix play a direct role in the dynamics of late endosome internal membranes in vivo. Endocytosed anti-LBPA antibodies interfere with protein and lipid sorting and trafficking, membrane transport and motility at the level of late endosomes, and these defects phenocopy the cholesterol storage disorder Niemann-Pick type C or NPC.[47-49] Presumably, antibodies, by binding their antigen within the endosomal lumen,[26] inhibit the dynamic properties of this intraluminal membrane system, and thus prevent the movement of proteins and lipids from intraluminal vesicles to the limiting membrane. Consistently, cholesterol accumulation within late endosomes, including in NPC cells, recapitulates the same defects as observed with anti-LBPA antibodies,[1] presumably because excess cholesterol, beyond the endosomal capacity, also collapses the dynamics of internal membrane and causes an endosomal traffic jam.[50] These observations lead to the notion that intraluminal proteins and lipids can be delivered to the limiting membrane of the organelle, by back-fusion of intraluminal vesicles with the limiting membrane, similarly to the transport of MHC class II molecules in antigen-presenting cells[12]—a process inhibited by anti-LBPA antibodies and sensitive to excess cholesterol.

Further evidence supporting this notion comes from studies with anthrax toxin and with the enveloped virus vesicular stomatitis virus (VSV). The protective antigen (PA) of anthrax toxin binds to a cell surface receptor, undergoes heptamerization, and then recruits the enzymatic subunits, the lethal factor (LF) and the edema factor (EF). After endocytosis of the complex, and membrane insertion of PA, LF and EF are ultimately delivered to the cytoplasm where their targets reside. Recent studies show that membrane insertion of PA already occurs in early endosomes, possibly only in the multivesicular regions, but that subsequent delivery of LF to the cytoplasm occurs preferentially later in the endocytic pathway, relies on the dynamics

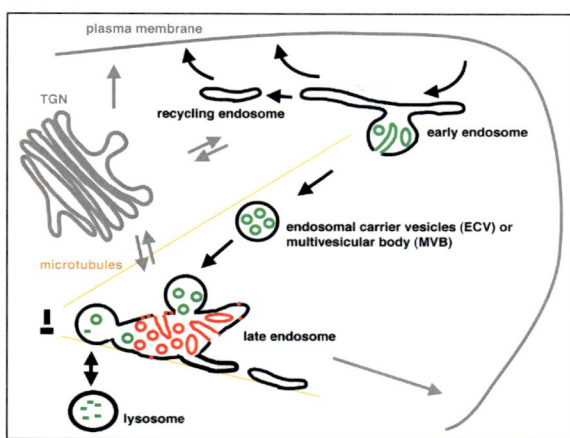

Figure 1. See page 20 for figure legend.

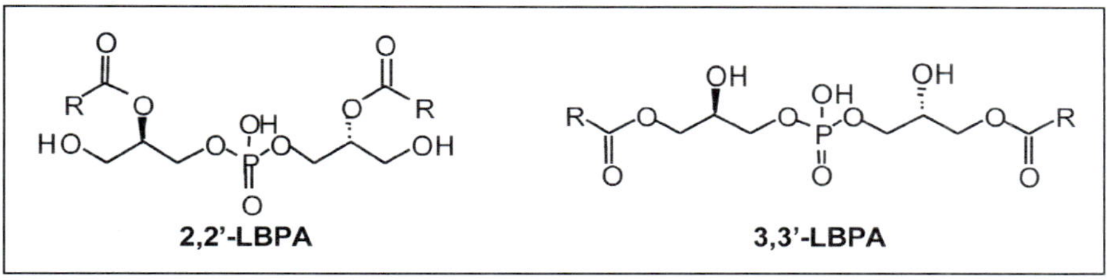

Figure 2. See page 20 for figure legend.

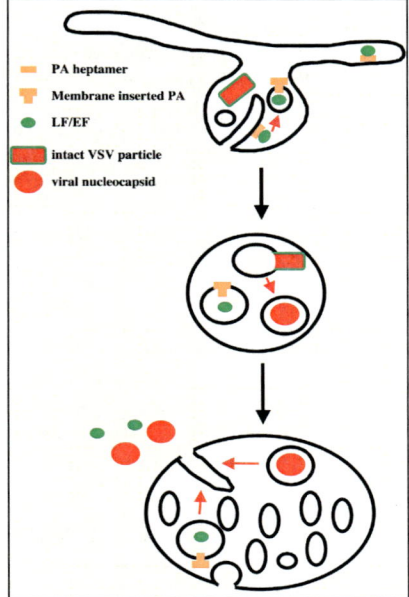

Figure 3. See page 20 for figure legend.

Figure 1. Endosomes and internal membranes. The figure shows the outline of the endocytic pathway in animal cells, as well as recycling routes to the trans-Golgi network or TGN (followed for example by MPR molecules), and to the plasma membrane (e.g., MHC class II in antigen-presenting cells). The two types of intraluminal membranes discussed in the text are indicated: (1) Green membranes contain down-regulated signalling receptors and may be enriched in PI3P and perhaps cholesterol. These vesicles and their cargoes are presumably transport to lysosomes for degradation. (2) Red membranes contain LBPA and may regulate the dynamics of late endosomes internal membranes via fission from and fusion with the limiting membrane. Much like at other transport steps in the cell,[53] two views have been proposed to account for endosome biogenesis, through their formation from a stable early endosome or through the maturation of early endosomes. These two alternative models have already been discussed extensively elsewhere, and there is still no experimental evidence that proves or discloses either model. It is not easy to discriminate between a multivesicular endosome that forms after the detachment of recycling tubules and one that forms as a result of direct detachment from a "stable" early endosome.

Figure 2. Structure of LBPA. The chemical composition of 2,2'-LBPA and 3,3'-LBPA is shown. In the BHK cells, the major (≈ 90%) isoform is 2,2'-dioleoyl-LBPA (R is C18:1). LBPA esterified at the 2 and 2' of the glycerol backbone is predicted to be cone-shaped and has the capacity to deform bilayers,[44] consistently with the role of lipid shapes in generating membrane curvature.[6]

Figure 3. Dynamics of intraluminal membranes. The figure outlines the intra-endosomal routes of anthrax toxin and VSV. After internalization, the protective antigen (PA) with bound lethal factor and edema factor (EF) inserts into the membrane of internal vesicles, presumably in early endosomes, and thereby allows the translocation of the LF and EF into the lumen of internal vesicles. These are transported to late endosomes. LF and EF are then released into the cyto-plasm via fusion of internal vesicles with the limiting membrane. Similarly, after endocytosis, intact VSV particles appear in early endosomes. These virions then fuse with intra-endosomal vesicles presumably within ECV/MVBs (see Fig. 1). Internal vesicles are transported to late endosomes, where the nucleocapsids are then released into the cytoplasm via fusion of inter-nal vesicles with the limiting membrane.

of internal vesicles of multivesicular late endosomes, and, in particular, is inhibited by Alix downexpression with siRNAs[51] (Fig. 3). Similarly, after VSV endocytosis and beyond early endosomes, the low endosomal pH triggers the fusion of the viral envelope with endosomal membranes, releasing the viral nucleocapsid into the cytosol, where replication of the viral genome occurs. Much like intoxication with anthrax, VSV infection is inhibited by Alix siRNAs[44] (Fig. 2). Consistently, recent studies, indicate that viral fusion already occurs in transport intermediates between early and late endosomes, presumably releasing the nucleocapsid within the lumen of intra-endosomal vesicles, where it remains hidden. Transport to late endosomes is then required for the nucleocapsid to be delivered to the cytoplasm, in a process that depends on LBPA and Alix.[52] Hence, it thus seems that VSV and anthrax toxin have hijacked the same mechanism for infection and intoxication, via the back-fusion of intralumi-nal vesicles with the late endosome limiting membrane, to overcome the diffusion barrier im-posed by the cortical actin cytoskeleton, and reach more efficiently the perinuclear region of the cell (Fig. 3).

Conclusions

The molecular events that regulate fission and fusion within the endosomal lumen clearly remain to be elucidated. It is far from clear how fission and fusion can be controlled by cytosolic machineries from the opposite side of the membrane—as opposed to the role of coat proteins and SNAREs in intracellular transport. Within late endosomes, our previously published data argue that these intraluminal fission and fusion events may depend, at least in part, on the intrinsic properties of the bilayer itself, via LBPA.[27,44] LBPA-rich membranes may have a high propensity to interact spontaneously with the limiting bilayer, involving some sort of kiss-and-run fission and fusion events. However, proteins are likely to control the process, since, in particular, fission may remain frustrated if uncontrolled fusion occurs concomitantly, and vice-versa. Our previous data suggested that Alix negatively controls the invagination process by binding LBPA-rich membranes,[44] and in turn, Alix is likely to act in concert with other proteins, including ESCRTs.[5] Thus, a simple and naïve view is that Alix traps the membrane intermediate in fission or fusion, by interacting with LBPA as the lipid appears on the limiting bilayer, and thereby controls the rates of both vesicle formation (invagination) and consumption (fusion). This view is attractive, because it provides a reasonable mechanistic explanation for the coupling, which must exist between invagination and back-fusion. Indeed, if uncoupled, the internal membrane pool would disappear (uncontrolled back–fusion), or the organelle collapse (uncontrolled fission).

Acknowledgments

We also wish to thank Gisou van der Goot and Pierre Luyet for critical reading of the manuscript. The Swiss National Science Foundation and the International Human Frontier Science Program supported the work described here.

References

1. Gruenberg J. The endocytic pathway: A mosaic of domains. Nat Rev Mol Cell Biol 2001; 2(10):721-30.
2. Zerial M, McBride H. Rab proteins as membrane organizers. Nat Rev Mol Cell Biol 2001; 2(2):107-17.
3. Simonsen A, Wurmser AE, Emr SD et al. The role of phosphoinositides in membrane transport. Curr Opin Cell Biol 2001; 13(4):485-92.
4. Gruenberg J. Lipids in endocytic membrane transport and sorting. Curr Opin Cell Biol 2003; 15(4):382-8.
5. Dikic I. ALIX-ing phospholipids with endosome biogenesis. Bioessays 2004; 26(6):604-7.
6. Huttner WB, Zimmerberg J. Implications of lipid microdomains for membrane curvature, budding and fission. Curr Opin Cell Biol 2001; 13(4):478-84.
7. van Meer G, Sprong H. Membrane lipids and vesicular traffic. Curr Opin Cell Biol 2004; 16(4):373-8.
8. Mukherjee S, Maxfield FR. Membrane domains. Annu Rev Cell Dev Biol 2004; 20:839-66.
9. Parton RG, Richards AA. Lipid rafts and caveolae as portals for endocytosis: New insights and common mechanisms. Traffic 2003; 4(11):724-38.
10. Simons K, Vaz WL. Model systems, lipid rafts, and cell membranes. Annu Rev Biophys Biomol Struct 2004; 33:269-95.
11. Lafont F, Abrami L, van der Goot FG. Bacterial subversion of lipid rafts. Curr Opin Microbiol 2004; 7(1):4-10.
12. Murk JL, Stoorvogel W, Kleijmeer MJ et al. The plasticity of multivesicular bodies and the regulation of antigen presentation. Semin Cell Dev Biol 2002; 13(4):303-11.
13. Kobayashi T, Gu F, Gruenberg J. Lipids, lipid domains and lipid-protein interactions in endocytic membrane traffic. Semin Cell Dev Biol 1998; 9(5):517-26.
14. Sandvig K, van Deurs B. Endocytosis, intracellular transport, and cytotoxic action of Shiga toxin and ricin. Physiol Rev 1996; 76(4):949-66.
15. Pagano RE. Lipid traffic in eukaryotic cells: Mechanisms for intracellular transport and organelle-specific enrichment of lipids. Curr Opin Cell Biol 1990; 2(4):652-63.
16. Marks DL, Pagano RE. Endocytosis and sorting of glycosphingolipids in sphingolipid storage disease. Trends Cell Biol 2002; 12(12):605-13.
17. Worby CA, Dixon JE. Sorting out the cellular functions of sorting nexins. Nat Rev Mol Cell Biol 2002; 3(12):919-31.
18. Carlton J, Bujny M, Rutherford A et al. Sorting nexins—unifying trends and new perspectives. Traffic 2005; 6(2):75-82.
19. Gillooly DJ, Morrow IC, Lindsay M et al. Localization of phosphatidylinositol 3-phosphate in yeast and mammalian cells. EMBO J 2000; 19(17):4577-88.
20. Odorizzi G, Babst M, Emr SD. Fab1p PtdIns(3)P 5-kinase function essential for protein sorting in the multivesicular body. Cell 1998; 95(6):847-858.
21. Ikonomov OC, Sbrissa D, Shisheva A. Mammalian cell morphology and endocytic membrane homeostasis require enzymatically active phosphoinositide 5-kinase PIKfyve. J Biol Chem 2001; 276(28):26141-7.
22. Friant S, Pecheur EI, Eugster A et al. Ent3p is a PtdIns(3,5)P2 effector required for protein sorting to the multivesicular body. Dev Cell 2003; 5(3):499-511.
23. Eugster A, Pecheur EI, Michel F et al. Ent5p is required with Ent3p and Vps27p for ubiquitin-dependent protein sorting into the multivesicular body. Mol Biol Cell 2004; 15(7):3031-41.

24. Dove SK, Piper RC, McEwen RK et al. Svp1p defines a family of phosphatidylinositol 3,5-bisphosphate effectors. EMBO J 2004; 23(9):1922-33.
25. Mobius W, van Donselaar E, Ohno-Iwashita Y et al. Recycling compartments and the internal vesicles of multivesicular bodies harbor most of the cholesterol found in the endocytic pathway. Traffic 2003; 4(4):222-31.
26. Kobayashi T, Stang E, Fang KS et al. A lipid associated with the antiphospholipid syndrome regulates endosome structure and function. Nature 1998; 392(6672):193-7.
27. Kobayashi T, Beuchat MH, Chevallier J et al. Separation and characterization of late endosomal membrane domains. J Biol Chem 2002; 277(35):32157-64.
28. Fivaz M, Vilbois F, Thurnheer S et al. Differential sorting and fate of endocytosed GPI-anchored proteins. EMBO J 2002; 21(15):3989-4000.
29. Sobo K, Chevallier J, Le Blanc I et al. Limiting and internal membranes contain raft domains with different biochemical properties and protein composition, (submitted).
30. Gruenberg J, Stenmark H. The biogenesis of multivesicular endosomes. Nat Rev Mol Cell Biol 2004; 5(4):317-23.
31. Bache KG, Slagsvold T, Stenmark H. Defective downregulation of receptor tyrosine kinases in cancer. EMBO J 2004; 23(14):2707-12.
32. Hicke L. Protein regulation by monoubiquitin. Nat Rev Mol Cell Biol 2001; 2(3):195-201.
33. Katzmann DJ, Odorizzi G, Emr SD. Receptor downregulation and multivesicular-body sorting. Nat Rev Mol Cell Biol 2002; 3(12):893-905.
34. Raiborg C, Rusten TE, Stenmark H. Protein sorting into multivesicular endosomes. Curr Opin Cell Biol 2003; 15(4):446-55.
35. Lloyd TE, Atkinson R, Wu MN et al. Hrs regulates endosome membrane invagination and tyrosine kinase receptor signaling in Drosophila. Cell 2002; 108(2):261-269.
36. Bache KG, Brech A, Mehlum A et al. Hrs regulates multivesicular body formation via ESCRT recruitment to endosomes. J. Cell Biol 2003; 162(3):435-442.
37. Brotherus J, Renkonen O, Herrmann J et al. Novel stereochemical configuration in lysobisphosphatidic acid of cultured BHK cells. Chem Phys Lipids 1974; 13:178-182.
38. Escola JM, Kleijmeer MJ, Stoorvogel W et al. Selective enrichment of tetraspan proteins on the internal vesicles of multivesicular endosomes and on exosomes secreted by human B- lymphocytes. J Biol Chem 1998; 273(32):20121-7.
39. Chow A, Toomre D, Garrett W et al. Dendritic cell maturation triggers retrograde MHC class II transport from lysosomes to the plasma membrane. Nature 2002; 418(6901):988-94.
40. Griffiths G, Hoflack B, Simons K et al. The mannose-6-phosphate receptor and the biogenesis of lysosomes. Cell 1988; 52:329-341.
41. Prescianotto-Baschong C, Riezman H. Ordering of compartments in the yeast endocytic pathway. Traffic 2002; 3(1):37-49.
42. Amidon B, Brown A, Waite M. Transacylase and phospholipases in the synthesis of bis(monoacylglycero)phosphate. Biochemistry 1996; 35:13995-14002.
43. Chevallier J, Sakai N, Robert F et al. Rapid access to synthetic lysobisphosphatidic acids using P(III) chemistry. Org Lett 2000; 2(13):1859-61.
44. Matsuo H, Chevallier J, Mayran N et al. Role of LBPA and Alix in multivesicular liposome formation and endosome organization. Science 2004; 303:531-534.
45. von Schwedler UK, Stuchell M, Müller B et al. The protein network of HIV budding. Cell 2003; 114:701-713.
46. Strack B, Calistri A, Craig S et al. AIP1/Alix is a binding partner for HIV-1 p6 and EIAV p9 functioning in virus budding. Cell 2003; 114:689-699.
47. Kobayashi T, Beuchat MH, Lindsay M et al. Late endosomal membranes rich in lysobisphosphatidic acid regulate cholesterol transport. Nat Cell Biol 1999; 1(2):113-8.
48. Lebrand C, Corti M, Goodson H et al. Late endosome motility depends on lipids via the small GTPase Rab7. EMBO J 2002; 21(6):1289-300.
49. Mayran M, Parton RG, Gruenberg J. Annexin II regulates multivesicular endosome biogenesis in the degradation pathway of animal cells. EMBO J 2003; 13:3242-3253.
50. Simons K, Gruenberg J. Jamming the endosomal system: Lipid rafts and lysosomal storage diseases. Trends Cell Biol 2000; 10(11):459-62.
51. Abrami L, Lindsay M, Parton RG et al. Membrane insertion of anthrax protective antigen and cytoplasmic delivery of lethal factor occur at different stages of the endocytic pathway. J Cell Biol 2004; 166(5):645-51.
52. Le Blanc I, Luyet P-P, Pons V et al. Endosome-to-cytosol transport of viral nucleocapsids. Nature Cell Biology 2005; 7(7):653-64.
53. Pelham HR, Rothman JE. The debate about transport in the Golgi—two sides of the same coin? Cell 2000; 102(6):713-9.

CHAPTER 3

Rab Domains on Endosomes

Marta Miaczynska* and Marino Zerial

Abstract

Small GTPases of the Rab family have been long recognized to be key regulators of membrane trafficking. However, recent studies have uncovered their more fundamental role as determinants of organelle biogenesis and maintenance in all cells. Rab proteins acting in the endocytic pathway were shown to occupy nonoverlapping, morphologically and biochemically distinct domains on membranes of endosomes. Molecular characterization of Rab5 and its effectors revealed basic principles by which this GTPase mediates local changes in membrane structure and function, thus organizing a specific domain on early endosomes. Rab domains on endosomes appear to coordinate multiple functions related to membrane trafficking, organelle motility and signal transduction and are dynamically linked through the activity of bivalent Rab effectors. The concept of Rab proteins acting as membrane organizers provides a framework explaining the biogenesis of endocytic organelles composed of separate but functionally coupled domains which are arranged in a dynamic fashion.

Introduction

Eukaryotic cells are characterized by highly compartmentalized structure comprising numerous membrane-bound organelles, which ensure a precise spatial segregation and temporal control of various physiological processes. Throughout evolution, polarization of cells and their functional specialization into tissues have been accompanied by changes in their intracellular organization, often resulting in specialized organelles present only in certain cell types, such as apical and basolateral endosomes in epithelial cells, melanosomes in pigment cells or dense-core granules in various secretory cells. The overall morphology and function of intracellular compartments have been investigated intensively for a few decades. However, studies of the sub-structure and the organization of membranes limiting the intracellular compartments have become possible only more recently owing to the developments of experimental techniques. A general concept emerging from studies at the sub-organellar level reveals that components constituting a membrane of a compartment, both proteins and lipids, are not stochastically distributed but rather segregated and concentrated in distinct but dynamic domains within the plane of the membrane. Importantly, this further implies that various functions within an organelle can be efficiently compartmentalized and assigned to appropriate domains. Finally, the identity of an organelle will be therefore determined by a particular combination of functional domains, which ensure a spatio-temporal regulation of processes taking place within this organelle.

Various mechanisms, based on protein-protein, protein-lipid and lipid-lipid interactions, appear to be responsible for formation of membrane domains. A classical example of local accumulation of specific proteins and lipids is the formation of coat complexes, required for

*Corresponding Author: Marta Miaczynska—International Institute of Molecular and Cell Biology, Ks. Trojdena 4, 02-109 Warsaw, Poland. Email: miaczynska@iimcb.gov.pl

Endosomes, edited by Ivan Dikic. ©2006 Landes Bioscience and Springer Science+Business Media.

concentration of selected cargo molecules and the budding of transport vesicles.[1] Coat components ensure generation of curved membrane pits, sorting and incorporation of cargo into them and finally their scission from the membrane. This series of events is orchestrated by an intricate network of interactions between proteins and lipids, occurring in a spatially confined membrane domain with a strict temporal control. Among various lipid classes involved, phosphoinositides seem to play a particularly important role in recruiting cytosolic proteins to the membrane.[2] Interestingly, also certain proteins seem to have a peculiar ability to drive the formation of specialized membrane structures, such as caveolin or von Willebrand factor which, again through further interactions with other proteins and lipids, direct biogenesis of caveolae[3] or Weibel-Palade bodies,[4] respectively. Interactions between lipids can also underlie a nonhomogenous distribution of membrane components, as exemplified by lipid rafts. Lipid rafts are generated through interactions of sphingolipids and cholesterol and selectively incorporate certain transmembrane proteins, fulfilling important regulatory functions.[5] In case of all membrane domains, local concentration of components, kept in place by mutual interactions, is crucial for confining specific functions to certain membrane regions.

This chapter will be devoted to another group of proteins involved in organelle biogenesis and postulated to form membrane domains that are small GTPases of the Rab family. In particular, we will discuss the mechanisms by which Rab proteins orchestrate intracellular transport via the spatio-temporal regulation of effector proteins that assemble into biochemically distinct and functionally specialized membrane domains on endosomal organelles.

Rab Proteins As Determinants of Organelle Identity

The Rab family of proteins comprises over 60 members (designated Ypt proteins in yeast) which regulate virtually all membrane trafficking steps within the secretory and endocytic pathways. They coordinate subsequent stages of transport such as formation of vesicles, their motility along cytoskeletal filaments and finally their docking and fusion with target membranes (reviewed in refs. 6,7). Newly synthesized, GDP-bound Rab proteins form a cytosolic complex with a Rab escort protein (REP) which presents them to geranylgeranyl transferase II for prenylation and subsequently delivers the modified proteins to their target membranes.[8,9] Following this initial, REP-mediated membrane targeting event, the activity of Rab proteins is regulated by two overlapping cycles (reviewed in refs. 10,11). First, Rab proteins locked in an inactive, GDP-bound form can shuttle between the cytosol and specific target membranes, chaperoned by Rab GDP dissociation inhibitor (GDI).[12] GDI is structurally related to REP; however it cannot mediate the prenylation of Rab proteins.[13] Second, once delivered to the membrane via a GDI displacement factor (GDF),[14] Rab proteins undergo cycles of activation resulting from binding of GTP, followed by inactivation via GTP hydrolysis. The nucleotide cycle of each Rab protein is catalyzed by specific GDP/GTP-exchange factors (GEFs) and GTPase activating proteins (GAPs). Active Rab proteins present on the membrane interact with their specific effectors, mediating downstream processes such as budding, motility or fusion of vesicles. Due to the regulated cycles of GTP binding and hydrolysis, followed by binding of effectors, Rab proteins ensure temporal and spatial control of membrane transport.[15]

Each Rab protein is characterized by a specific and restricted intracellular distribution. Several Rabs have been localized to endosomal compartments and/or implicated in the regulation of various endocytic events. At present, this list includes: Rab4,[16] Rab5,[17] Rab7,[18] Rab9,[19] Rab11,[20] Rab13,[21] Rab14,[22] Rab15,[23] Rab17,[24,25] Rab18,[26] Rab20,[26] Rab21,[27] Rab22,[28] Rab23,[29] Rab25,[30] Rab34[31] and Rab39.[32] However, only a few of these proteins have been characterized in more detail. Among them, the ubiquitously expressed Rab4, Rab11 and Rab15 are present on early and recycling endosomes,[20,23,33-35] although a Golgi-associated pool of Rab11 also exists.[36] Rab5 localizes to clathrin-coated vesicles and early endosomes,[17] Rab22 is present on early endosomes,[28,37] while Rab7 and Rab9 are distributed to late endosomes.[18,19] Some Rab proteins such as Rab17,18,20 or 25 are specifically expressed in epithelial cells where Rab17 and 25 appear to regulate polarized endocytosis through the apical recycling compartment.[25,30]

Due to their specific localization to various membrane compartments throughout the cell,[6] Rab proteins have been long recognized as organelle markers and their role as rate-limiting regulators of transport is well established.[15,17,38,39] However, more recent data using knockout/knockdown approaches argue that Rab proteins are not merely "compartment tags" but play an active role in the biogenesis of membrane organelles, being one of the key determinants of compartment identity. Two studies have recently provided strong experimental evidence in support of this concept. In *Drosophila*, zygotic loss of Rab5 causes drastic disruption of endosomes during initial stages of development and, eventually, leads to embryonic lethality.[40] Knockdown of Rab9 by RNA interference (RNAi) in cultured mammalian cells decreased the overall size of late endosomes.[41] Strikingly, it also reduced a number of particular subclasses of these endosomes, such as multilamellar and dense-tubule–containing late endosomes/lysosomes, but not multivesicular endosomes. These data strongly argue that Rab proteins play a crucial role in the biogenesis of endocytic organelles.

Rab Proteins As Organizers of Membrane Domains

Are Rab proteins evenly distributed throughout the organelle membrane? Some initial observations indicated that Rab5 was not present uniformly on the membrane of early endosomes but rather concentrated in clusters.[42,43] Such assemblies were visualized by light microscopy on endosomes enlarged due to the overexpression of an activated mutant of Rab5 (Rab5Q79L). Interestingly, Rab5-enriched clusters contained also a Rab5 effector EEA1[42] and were concentrated in regions mediating fusion between endosomes.[43] Similarly, docking of yeast vacuoles before fusion appears to involve formation of "vertex" ring-shaped microdomains around the periphery of the apposed membranes. These vertices are selectively enriched in Rab GTPase Ypt7p together with its effector complex Vps class C/HOPS.[44] Thus, Rab proteins with the associated effector proteins appear to mark particular regions of the organelle membrane, thus predestining them for certain functions.

However, most organelles appear to contain more then one Rab protein, raising questions about the distribution and any functional relationships between Rab proteins within the same membrane compartment. A systematic analysis of distribution of endosomal Rab5, Rab4 and Rab11 with respect to endocytic cargo (transferrin) has been conducted by quantitative light microscopy analysis.[34] Strikingly, the analyzed Rab proteins exhibited a largely nonoverlapping distribution, with each protein occupying distinct, often adjacent membrane regions within the individual endosomal compartments. While early endosomes appeared to be composed of domains containing Rab5 and Rab4, recycling endosomes represented a mosaic of Rab4 and Rab11 domains (Fig. 1). Such distribution was nonstochastic, as Rab5 was present in various amounts in different pools: about 50% of all Rab5 structures did not contain Rab4 or Rab11, 30% of them colocalized only with Rab4 and 20% contained both Rab4 and Rab11. In contrast, 30% of Rab4 compartments were positive for Rab5; other 30% contained Rab11, 20% colocalized with both Rab5 and Rab11, while the remaining 20% were Rab5- and Rab11-negative. Moreover, internalized transferrin (endocytic cargo destined for recycling) colocalized sequentially first with Rab5 domains, then with Rab4- and finally with Rab11-enriched regions. Similar functional segregation of Rab4 and Rab11 domains with respect to cargo has been demonstrated for recycling of glycosphingolipids using fluorescent analogue of lactosylceramide as a marker,[45] underscoring the notion that lipids and proteins segregate into different membrane (micro)domains. Further analyses revealed that domains containing distinct Rab proteins exhibited different pharmacological properties.[34,45] While domains on Rab4 and Rab11 endosomes appeared to be sensitive to brefeldin A (BFA), Rab5 domains were resistant to BFA-induced tubulation but instead affected by phosphatidylinositol 3-kinase (PI3K) inhibitor wortmannin. These data indicated that other components of endosomal membranes are also selectively concentrated in specific Rab domains.

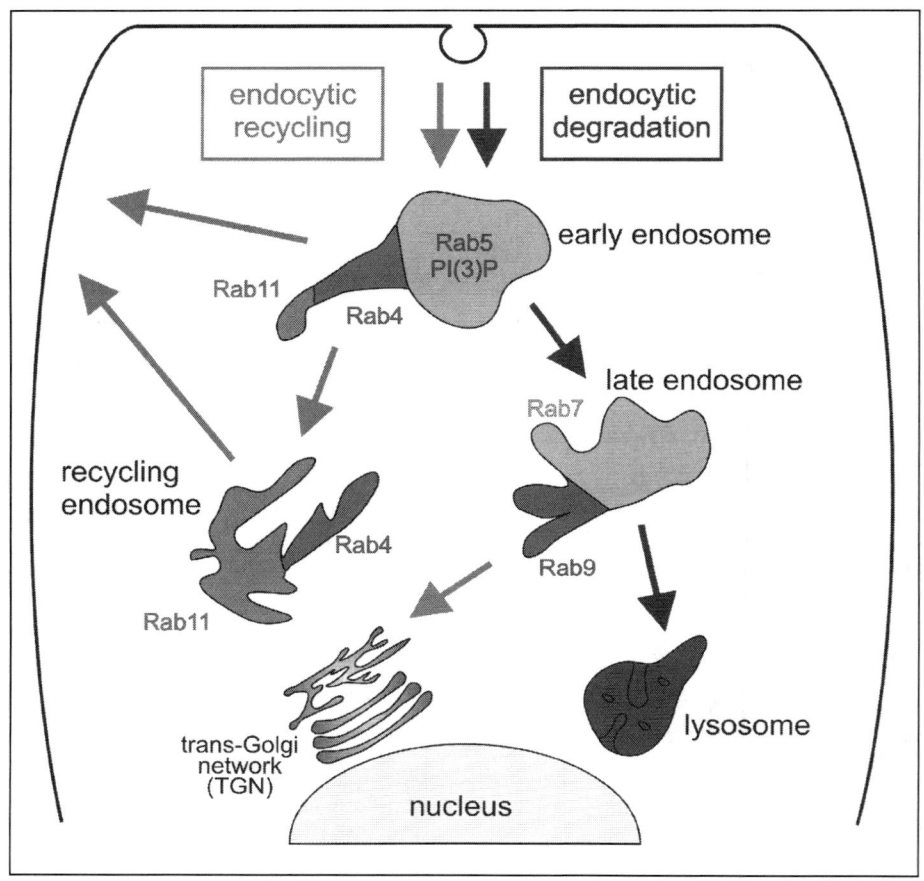

Figure 1. Distribution of Rab domains along the endocytic pathways. Only major endosomal compartments are shown. Cargo destined for degradation or recycling enters the early endosome via the Rab5 domain. Subsequently, cargo is sorted either for recycling via Rab4 and Rab11 domains or for degradation via Rab7 domains. Rab9 domains on late endosomes direct transport of certain cargo (e.g., CI-MPR) towards the trans-Golgi network.

A complementary study[46] demonstrated a similar principle of organization of late endosomes where Rab7 and Rab9 were shown to occupy distinct domains within a single organelle. Rab9 domains, regulating transport between late endosomes and trans-Golgi network (TGN), are enriched in specific cargo (cation-independent mannose 6-phosphate receptors, CI-MPR) and contain Rab9 effector TIP47. In contrast, Rab7 domains are postulated to mediate the transport of cargo from early endosomes towards degradation in lysosomes. Thus, two different trafficking routes through late endosomes appear to be spatially separated and regulated by distinct Rab domains. Overall, the endosomal compartments can therefore be considered as a mosaic of various domains occupied by Rab proteins and their effectors.

Molecular Assembly of a Rab Domain

A key feature of Rab proteins acting as organizers of membrane domains is their ability to mediate local changes in membrane structure and function. This is achieved via a series of interactions with a large number of protein effectors, among them lipid-modifying enzymes.

The latter can affect the lipid composition of the bilayer by local generation and accumulation of particular lipid species. In parallel, Rab-mediated recruitment of cytosolic proteins can locally modulate the protein content of the membrane, creating a microenvironment enriched in certain molecules and thus predestined for certain functions. Cooperativity of effector recruitment, membrane anchoring of effectors through binding to specific lipids and lateral interactions between recruited effectors constitute the major principles of domain formation and maintenance by Rab proteins. These principles are illustrated by the best-studied example of a domain coordinated by Rab5 on the membrane of early endosomes (Fig. 2).

Rab5, initially delivered to the endosomal membrane in an inactive, GDP-bound form, undergoes activation catalyzed by specific GEFs, such as Rabex-5,[47] RIN1,[48] RIN2[49] or RIN3.[50] Interestingly, Rabex-5 is stably complexed to a Rab5 effector Rabaptin-5.[47] Such physical association of a GEF and an effector ensures a synergistic action of both molecules.[51] On the one side, Rabaptin-5 increases the exchange activity of Rabex-5 on Rab5. On the other side, Rab5-dependent recruitment of Rabaptin-5 to early endosomes is completely dependent on its physical association with Rabex-5. Rab5 on early endosomes undergoes continuous cycles of nucleotide binding and hydrolysis,[15] the latter process assisted by specific GAP proteins such as RN-tre[52] or RabGAP-5.[53] Active Rab5 can further interact with other effectors, one of them being the type III PI3K complex hVPS34/p150.[54] Although this complex is targeted to early endosomes Rab5-independently,[55] most likely via lipid-modified p150,[56] an interaction between Rab5-GTP and p150 occurring on the membrane is believed to locally activate the PI3K and thus restrict the production of phosphatidylinositol 3-phosphate (PI(3)P) to a particular domain. The resulting accumulation of Rab5-GTP and PI(3)P creates high-affinity binding sites for recruitment of cytosolic effectors such as EEA1,[57] Rabenosyn-5[58] or Rabankyrin-5[59] able to bind both PI(3)P via a FYVE motif (named after Fab1p, YOTB, Vac1p and EEA1[60]) and Rab5 (Fig. 2). Additionally, lateral interactions between the recruited effectors and other membrane components, such as the elements of the SNARE [soluble N-ethylmaleimide-sensitive factor attachment protein (SNAP) receptor] machinery, lead to formation of large oligomers,[42] stabilizing the molecular backbone of the domain. Indeed, experiments using fluorescence recovery after photobleaching (FRAP) on individual Rab5Q79L-enlarged endosomes demonstrated a restricted lateral mobility of GFP-Rab5Q79L molecules, consistent with the existence of oligomeric effector complexes on the early endosome membrane.[61] Recruitment of membrane tethering/fusion complexes is further coupled to the cytoskeletal transport machinery. The plus-end kinesin KIF16B is recruited to early endosomes Rab5- and PI(3)P-dependently and is rate-limiting for the association and movement of early endosomes with microtubules [62] (see below).

In addition to its interaction with hVPS34/p150, the type III PI3K, on the early endosomes, Rab5 binds to and stimulates the activity of p110β/p85α, the type I PI3K which is recruited to the plasma membrane in response to growth factor or cytokine stimulation and Ras activation [54,63,64] (Fig. 2). This type of PI3K converts PI(4,5)P$_2$ and PI(4)P to PI(3,4,5)$_3$ and PI(3,4)P$_2$, respectively, which in turn play an important role in growth factor signaling, actin rearrangements, phagocytosis and cell motility by recruiting appropriate effector proteins to the plasma membrane.[63] Indeed, Rab5 has been previously implicated in the regulation of cell motility, [65] phagocytosis [66-68] and various aspects of growth factor signaling [48,69-71] (see below). Surprisingly, Rab5 appears to coordinate not only the production of the 3-phosphorylated inosites but also their turnover through interactions with the specific phosphatases.[64] Rab5 binds directly and stimulates the activity of the type II inositol 5-phosphatase and the type I α PI(3,4)P$_2$ 4-phosphatase, thus promoting the gradual dephosphorylation of PI(3,4,5)$_3$ generated at the plasma membrane to PI(3)P accumulating on early endosomes (Fig. 2). These recent data provided first evidence for a Rab protein regulating both generation and turnover of phosphoinositides through an enzymatic cascade of effectors.

Other Rab proteins also appear to be functionally linked to different lipid-modifying enzymes, even though the molecular mechanisms of such interactions have not been explored

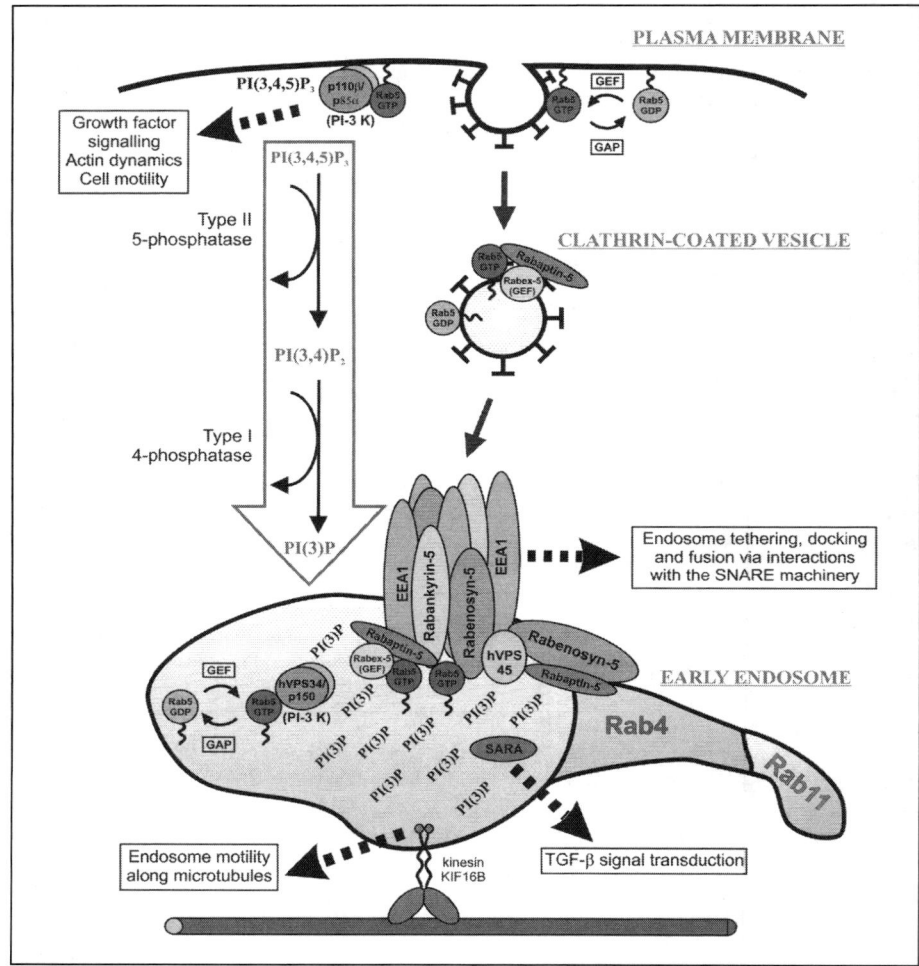

Figure 2. Interactions between Rab5 and its effectors in the endocytic pathway. Rab5 interacts with different effectors on the plasma membrane, clathrin-coated vesicles and early endosomes. The enzymatic cascade of Rab5 effectors which leads from PI(3,4,5)P$_3$ at the plasma membrane to PI(3)P at the early endosomes[64] is depicted by a red arrow. The molecular assembly of Rab5-PI(3)P domain on early endosome and its various functions are presented. See text for the detailed description.

in detail yet. PIKfyve, a protein and lipid kinase regulating the morphology of late endosomes through production of PI(5)P and PI(3,5)P$_2$[72] has been shown to interact with and phosphorylate the Rab9 effector p40.[73] As Rab9 regulates trafficking of cargo such as CI-MPR from late endosomes to TGN, the PIKfyve-p40 interaction has been proposed to function in this transport step, again indicating molecular links between the production of specific phosphoinositides and the Rab machinery. Similarly, phosphatidylinositol 4-kinase β (PI4Kβ) has been shown to interact with the active form of Rab11.[74] However, as Rab11 is present not only within the endosomal system but also in the Golgi complex,[36] this interaction appears to be important for the biosynthetic transport from the Golgi to the plasma membrane. Similarly, Golgi-resident PI4K Pik1p in yeast appears to be functionally linked to Rab11-related GTPase Ypt31p in regulating protein trafficking through the secretory pathway, although a direct interaction between the two proteins has not been demonstrated.[75]

Functional connections between lipid-metabolizing enzymes and Rab proteins represent a more general phenomenon observed for other subfamilies of small GTPases which use such enzymes as downstream effectors. Examples include PI3K as Ras effector,[76] phospholipase C-β2 as an effector of Rac[77] or phosphatidylinositol 4-phosphate 5-kinases as ARF6 and ARF1 effectors.[78,79] Although the domain distribution of these GTPases on their target membranes has not been systematically analyzed, it is very likely that their ability to affect the lipid composition of the bilayer leads to the formation of specific domains in a manner analogous to Rab proteins.

Functional Coordination between Components of the Rab5 Domain

The function(s) of Rab proteins are determined by the number and type of interacting with them effectors. Accordingly, single Rab protein capable of binding various effectors can mediate several functions. Rab5 shown to interact with over 30 proteins provides an example of a multifunctional GTPase, regulating several aspects of endocytic membrane transport.[30,81] Initially, over a decade ago, Rab5 has been characterized as a factor required for homotypic fusion of early endosomes and heterotypic fusion of early endosomes with plasma membrane-derived clathrin-coated vesicles.[17,82] A more recent identification of Rab5 effectors[80,81] has not only shed light on the molecular mechanisms of these processes but also uncovered new unexpected functions for Rab5, such as regulation of endosome motility or signal transduction (Fig. 2, see also below).

The involvement of Rab5 in endosomal fusion can be primarily attributed to the regulation of docking, a process preceding SNARE-mediated fusion reaction. Rab5 domains on early endosomes enriched in PI(3)P recruit EEA1 which is a crucial factor for endosome tethering and docking.[80] Moreover, EEA1 appears to form oligomeric complexes with the components of the SNARE machinery such as syntaxin 13, NSF or α-SNAP, thus most likely providing coupling between docking and fusion steps.[42] A second link with SNAREs is provided by Rabenosyn-5, another Rab5 effector recruited to the PI(3)P-enriched domain via its FYVE domain. Rabenosyn-5 interacts with a Sec1-like protein Vps45, which binds several endosomal syntaxins.[58] Interestingly, Rab5-PI(3)P domain on early endosomes seems to act as a docking platform for fusion of incoming clathrin-coated vesicles which contain Rab5 but are devoid of VPS34/p150 activity[54,83] (Fig. 2). Thus, Rab5 domain on early endosomes has an additional role in specifying the directionality of membrane transport from the plasma membrane. Moreover, excessive activation of Rab5 can direct caveolar vesicles, normally following an independent trafficking route, to fuse with early endosomes,[61] arguing that Rab5 controls also transport between caveolae/caveosomes and endosomes.

In addition to the regulation of docking and fusion, the presence of PI(3)P in Rab5 domains is crucial for other processes. One of them is motility of early endosomes along microtubules.[84] Recently, KIF16B kinesin motor containing PI(3)P-binding PX motif has been shown to mediate a Rab5-dependent, plus end-directed movement of early endosomes.[62] Deregulation of KIF16B and the resulting repositioning of early endosomes in the cell significantly affected the transport of endocytic cargo towards degradation or recycling. Thus, Rab5- and PI(3)P-dependent motility of endosomes appears to be crucial for proper endocytic trafficking.

Another process requiring the presence of PI(3)P on early endosomes is signal transduction in response to transforming growth factor-β (TGF-β) mediated by a FYVE-domain protein SARA (Smad anchor for receptor activation).[85] The localization of SARA to endosomes, which depends on PI(3)P and can be disrupted by dominant-negative Rab5 mutant,[86] is required for downstream signaling events, such as Smad2 nuclear translocation.[87] Interestingly, in addition to being one of the key components of TGF-β signaling cascade, SARA appears to have a second function as an endocytic factor, regulating the morphology of endosomes and transport of transferrin.[88] Such dual role of SARA may not be surprising in the light of tight mutual interdependence between endocytosis and signal transduction, many aspects of which are constantly being uncovered (for recent reviews see refs. 89-91).

Rab5 involvement in signal transduction is not limited to the recruitment of signaling proteins to endosomes, as Rab5 itself is a target of regulation by receptor tyrosine kinases (RTKs) at several levels. The enzymatic activity of Rab5 GEF RIN1 or GAP RN-tre can be modulated by growth factors such as epidermal growth factor (EGF), thus subjecting Rab5 nucleotide cycle and the resulting changes in endocytic rates to the regulation by signaling cascades.[48,52,69] Moreover, Rab5 seems to be required for RTK-induced actin remodeling in a process mediated by RN-tre.[92] A recent identification of APPL proteins as Rab5 effectors uncovered yet another role for Rab5 in transduction of signals from the plasma membrane to the nucleus.[71] APPL proteins are signal transducers required for cell proliferation, with a dual localization on endosomal membranes and in the nucleus. Interestingly, APPL-containing Rab5 endosomes appear to be distinct from the PI(3)P-positive compartments and preferentially accessible for certain cargo such as EGF but not transferrin, indicating that Rab5 may be involved in biogenesis of various endocytic structures besides canonical, PI(3)P-containing early endosomes.

Dynamic Coupling between Rab Domains

In order to regulate cargo transport along the endocytic routes, Rab domains need to be functionally linked. Transport of cargo to consecutive compartments, with an ultimate goal of degradation or recycling, appears to be achieved via sequential transfer between neighboring Rab domains (Fig. 1). As described above, transferrin is internalized into Rab5 domains and recycled passing through Rab4 and Rab11 domains.[34] At the molecular level, coupling between Rab domains is provided by two main mechanisms: 1) bivalent effectors, binding active forms of two Rab proteins, and 2) Rab-dependent recruitment of GEFs. The first example of a coupling protein was provided by Rabaptin-5, discovered to interact with both Rab5 and Rab4.[93] Subsequently, other cases of bi-functional effectors have been identified and include Rab4 and Rab5 effectors Rabenosyn-5[94] and Rabip4';[95] Rab5 and Rab22 effector EEA1;[37] Rab4 and Rab11 effector Rab Coupling Protein RCP[96] or Rab5 effector hVPS34/ p150[54] potentially interacting also with Rab7.[97] These molecules can regulate the morphology and functionality of Rab domains, arguing that coupling between them is not permanent but can be dynamically modulated. Indeed, overexpression of Rabenosyn-5 has been shown to increase the association between Rab5 and Rab4 endosomal domains, at the same time decreasing the fraction of Rab4- and Rab11-positive structures and resulting in a changed kinetics of transferrin recycling.[94] Although not directly demonstrated, it is plausible that hVPS34/p150, which generates PI(3)P, could link Rab5 and Rab7 domains to ensure transfer of cargo from early to late endosomes towards degradation. Interestingly, the presence of PI(3)P in Rab5 and Rab7 domains could result in recruitment of PIKfyve, which possesses itself a PI(3)P-binding motif and uses this lipid as a substrate for production of $PI(3,5)P_2$, characteristic of late endosomes.[72]

A second mechanism for sequential coupling of Rab domains has been uncovered by another study proposing that Rab proteins within yeast secretory pathway act in a cascade[98] with a preceding Rab recruiting a GEF to activate the consecutive one. In the described case, active Ypt31/32p present on the Golgi membranes bind their effector Sec2p which in turn acts as a GEF for another Rab GTPase, Sec4p.[99] While Ypt31/32p have been implicated in the intra-Golgi transport and budding of secretory vesicles from the Golgi membrane,[100,101] Sec4p regulates fusion of these vesicles with the plasma membrane.[102] By interaction with Ypt31/ 32p, Sec2p gets incorporated into secretory vesicles and ensures activation of Sec4p, which is required for vesicle exocytosis. An analogous mechanism could potentially act at various steps of endo- and exocytosis also in higher organisms, although further evidence in support of such model is currently lacking.

Future Prospects

The concept of Rab domains provides a framework explaining the organization of the endocytic organelles. However, several questions are posed by this model. In terms of molecular mechanisms, one of the key problems is the size and the temporal stability of Rab domains. They could represent relatively stable entities, able to dynamically grow or shrink but retaining a minimal steady 'core'. In such case, cargo would be sequentially transferred between the preexisting Rab domains. An alternative model envisages that Rab domains could be periodically disassembled and assembled de novo. In this option, consecutive Rab domains would be sequentially formed during transport of cargo on the membrane encapsulating it. Very recently, both models received some initial experimental support[103,104] but further careful quantitative imaging of endocytic transport in living cells should shed more light on the dynamics of Rab domains in time with respect to cargo flow. A related issue, which needs to be addressed as the visualization methods are improved in the future, is the size range of individual Rab domains and which consequences the domain size may have for the regulation of membrane flow.

Only a limited number of endocytic Rab proteins have been characterized in detail with respect to their exact intracellular localization, domain formation and interacting effectors, although potentially a large number of Rab proteins may regulate endocytosis. It is unclear at present whether all of them are able to actively form specialized membrane domains, like Rab5, or whether some of them only cosegregate within the already existing Rab domains. For example, it will be interesting to see whether Rab22 exhibiting the highest sequence homology to Rab5 and interacting with EEA1 similarly to Rab5[37] can form a separate domain on early endosomes. Extending this question, future studies should reveal whether Rab proteins expressed only in certain cell types are capable of forming specialized membrane domains or even whole specialized organelles in their target cells and whether they could do so also when introduced in a heterologous system. Systematic characterization of other Rab proteins will undoubtedly lead to a more complete picture of all Rab domains present in the cell and their mutual relationships.

Finally, it remains to be determined to what extent small GTPases from other families, such as Ras, Rho or Arf, act as organizers of membrane domains and how such domains could relate to Rab domains. Clearly, signals mediated by various GTPases, such as Ras-mediated signal transduction, Rho-dependent cytoskeleton rearrangements or Arf-regulated budding events, need to be functionally integrated with the membrane flow orchestrated by Rab proteins. Indeed, dual-specificity effectors binding GTPases of different classes have already been identified, such as Arfophilins which regulate the distribution of recycling endosomes and interact with Rab11 and Arf5[105] or the Exocyst complex binding a variety of GTPases of Rab, Rho, Ral and Arf subfamilies through its various components.[106-108] Further studies of small GTPases and their effectors will be pivotal for our understanding of how membrane compartmentalization into domains may specify the identity and function of intracellular organelles.

Acknowledgements

The work in Marino Zerial's laboratory was supported by grants from the HFSP (RG-0260/1999-M), the European Union (HPRN-CT-2000-00081) and the Max Planck Society.

References

1. McMahon HT, Mills IG. COP and clathrin-coated vesicle budding: Different pathways, common approaches. Curr Opin Cell Biol 2004; 16:379-391.
2. Parker PJ. The ubiquitous phosphoinositides. Biochem Soc Trans 2004; 32:893-898.
3. Parton RG. Caveolae—from ultrastructure to molecular mechanisms. Nat Rev Mol Cell Biol 2003; 4:162-167.
4. Michaux G, Cutler DF. How to roll an endothelial cigar: The biogenesis of Weibel-Palade bodies. Traffic 2004; 5:69-78.

5. Simons K, Ikonen E. Functional rafts in cell membranes. Nature 1997; 387:569-572.

6. Zerial M, McBride H. Rab proteins as membrane organizers. Nat Rev Mol Cell Biol 2001; 2:107-117.

7. Pfeffer SR. Rab GTPases: Specifying and deciphering organelle identity and function. Trends Cell Biol 2001; 11:487-491.

8. Andres DA, Seabra MC, Brown MS et al. cDNA cloning of component A of rab geranylgeranyl transferase and demonstration of its role as a rab escort protein. Cell 1993; 73:1091-1099.

9. Alexandrov K, Horiuchi H, Steele-Mortimer O et al. Rab escort protein-1 is a multifunctional protein that accompanies newly prenylated rab proteins to their target membranes. EMBO J 1994; 13:5262-5273.

10. Seabra MC, Wasmeier C. Controlling the location and activation of Rab GTPases. Curr Opin Cell Biol 2004; 16:451-457.

11. Pfeffer S, Aivazian D. Targeting Rab GTPases to distinct membrane compartments. Nat Rev Mol Cell Biol 2004; 5:886-896.

12. Sasaki T, Kikuchi A, Araki S et al. Purification and characterization from bovine brain cytosol of a protein that inhibits the dissociation of GDP from and the subsequent binding of GTP to smg p25A, a ras-like GTP-binding protein. J Biol Chem 1990; 265:2333-2337.

13. Schalk I, Zeng K, Wu S-K et al. Structure and mutational analysis of Rab GDP-disscociation inhibitor. Nature 1996; 381:42-48.

14. Sivars U, Aivazian D, Pfeffer SR. Yip3 catalyses the dissociation of endosomal Rab-GDI complexes. Nature 2003; 425:856-859.

15. Rybin V, Ullrich O, Rubino M et al. GTPase activity of rab5 acts as a timer for endocytic membrane fusion. Nature 1996; 383:266-269.

16. van der Sluijs P, Hull M, Webster P et al. The small GTP-binding protein rab4 controls an early sorting event on the endocytic pathway. Cell 1992; 70:729-740.

17. Bucci C, Parton RG, Mather IH et al. The small GTPase rab5 functions as a regulatory factor in the early endocytic pathway. Cell 1992; 70:715-728.

18. Feng Y, Press B, Wandinger-Ness A. Rab7: An important regulator of late endocytic membrane traffic. J Cell Biol 1995; 131:1435-1452.

19. Lombardi D, Soldati T, Riederer MA et al. Rab9 functions in transport between late endosomes and the trans Golgi network. EMBO J 1993; 12:677-682.

20. Ullrich O, Reinsch S, Urbé S et al. Rab11 regulates recycling through the pericentriolar recycling endosome. J Cell Biol 1996; 135:913-924.

21. Morimoto S, Nishimura N, Terai T et al. Rab13 mediates the continuous endocytic recycling of occludin to the cell surface. J Biol Chem 2005; 280:2220-2228.

22. Junutula JR, De Maziere AM, Peden AA et al. Rab14 is involved in membrane trafficking between the Golgi complex and endosomes. Mol Biol Cell 2004; 15:2218-2229.

23. Zuk PA, Elferink LA. Rab15 mediates an early endocytic event in Chinese hamster ovary cells. J Biol Chem 1999; 274:22303-22312.

24. Hunziker W, Peters PJ. Rab17 localizes to recycling endosomes and regulates receptor-mediated transcytosis in epithelial cells. J Biol Chem 1998; 273:15734-15741.

25. Zacchi P, Stenmark H, Parton RG et al. Rab17 regulates membrane trafficking through apical recycling endosomes in polarized epithelial cells. J Cell Biol 1998; 140:1039-1053.

26. Lütcke A, Valencia A, Olkkonen V et al. Cloning and subcellular localization of novel rab proteins reveals polarized and cell-type specific expression. J Cell Sci 1994; 107:3437-3448.

27. Simpson JC, Griffiths G, Wessling-Resnick M et al. A role for the small GTPase Rab21 in the early endocytic pathway. J Cell Sci 2004; 117:6297-6311.

28. Mesa R, Salomon C, Roggero M et al. Rab22a affects the morphology and function of the endocytic pathway. J Cell Sci 2001; 114:4041-4049.

29. Evans TM, Ferguson C, Wainwright BJ et al. Rab23, a negative regulator of hedgehog signaling, localizes to the plasma membrane and the endocytic pathway. Traffic 2003; 4:869-884.

30. Casanova JE, Wang X, Kumar R et al. Association of Rab25 and Rab11a with the apical recycling system of polarized Madin-Darby canine kidney cells. Mol Biol Cell 1999; 10:47-61.

31. Sun P, Yamamoto H, Suetsugu S et al. Small GTPase Rah/Rab34 is associated with membrane ruffles and macropinosomes and promotes macropinosome formation. J Biol Chem 2003; 278:4063-4071.

32. Chen T, Han Y, Yang M et al. Rab39, a novel Golgi-associated Rab GTPase from human dendritic cells involved in cellular endocytosis. Biochem Biophys Res Commun 2003; 303:1114-1120.

33. van der Sluijs P, Hull M, Zahraoui A et al. The small GTP-binding protein rab4 is associated with early endosomes. Proc Natl Acad Sci USA 1991; 88:6313-6317.

34. Sonnichsen B, De Renzis S, Nielsen E et al. Distinct membrane domains on endosomes in the recycling pathway visualized by multicolor imaging of Rab4, Rab5, and Rab11. J Cell Biol 2000; 149:901-914.
35. Zuk PA, Elferink LA. Rab15 differentially regulates early endocytic trafficking. J Biol Chem 2000; 275:26754-26764.
36. Urbé S, Huber LA, Zerial M et al. Rab11, a small GTPase associated with both constitutive and regulated secretory pathways in PC12 cells. FEBS Letters 1993; 334:175-182.
37. Kauppi M, Simonsen A, Bremnes B et al. The small GTPase Rab22 interacts with EEA1 and controls endosomal membrane trafficking. J Cell Sci 2002; 115:899-911.
38. Stenmark H, Parton RG, Steele-Mortimer O et al. Inhibition of rab5 GTPase activity stimulates membrane fusion in endocytosis. EMBO J 1994; 13:1287-1296.
39. Barbieri MA, Li G, Colombo M et al. Rab5, and early acting endosomal GTPase, supports in vitro endosome fusion without GTP hydrolysis. J Biol Chem 1994; 269:18720-18722.
40. Wucherpfennig T, Wilsch-Brauninger M, Gonzalez-Gaitan M. Role of Drosophila Rab5 during endosomal trafficking at the synapse and evoked neurotransmitter release. J Cell Biol 2003; 161:609-624.
41. Ganley IG, Carroll K, Bittova L et al. Rab9 GTPase regulates late endosome size and requires effector interaction for its stability. Mol Biol Cell 2004; 15:5420-5430.
42. McBride HM, Rybin V, Murphy C et al. Oligomeric complexes link Rab5 effectors with NSF and drive membrane fusion via interactions between EEA1 and syntaxin 13. Cell 1999; 98:377-386.
43. Roberts RL, Barbieri MA, Pryse KM et al. Endosome fusion in living cells overexpressing GFP-rab5. J Cell Sci 1999; 112:3667-3675.
44. Fratti RA, Jun Y, Merz AJ et al. Interdependent assembly of specific regulatory lipids and membrane fusion proteins into the vertex ring domain of docked vacuoles. J Cell Biol 2004; 167:1087-1098.
45. Choudhury A, Sharma DK, Marks DL et al. Elevated endosomal cholesterol levels in Niemann-Pick cells inhibit rab4 and perturb membrane recycling. Mol Biol Cell 2004; 15:4500-4511.
46. Barbero P, Bittova L, Pfeffer SR. Visualization of Rab9-mediated vesicle transport from endosomes to the trans-Golgi in living cells. J Cell Biol 2002; 156:511-518.
47. Horiuchi H, Lippé R, McBride HM et al. A novel Rab5 GDP/GTP exchange factor complexed to Rabaptin-5 links nucleotide exchange to effector recruitment and function. Cell 1997; 90:1149-1159.
48. Tall GG, Barbieri MA, Stahl PD et al. Ras-Activated Endocytosis Is Mediated by the Rab5 Guanine Nucleotide Exchange Activity of RIN1. Dev Cell 2001; 1:73-82.
49. Saito K, Murai J, Kajiho H et al. A novel binding protein composed of homophilic tetramer exhibits unique properties for the small GTPase Rab5. J Biol Chem 2002; 277:3412-3418.
50. Kajiho H, Saito K, Tsujita K et al. RIN3: A novel Rab5 GEF interacting with amphiphysin II involved in the early endocytic pathway. J Cell Sci 2003; 116:4159-4168.
51. Lippe R, Miaczynska M, Rybin V et al. Functional synergy between Rab5 effector Rabaptin-5 and exchange factor Rabex-5 when physically associated in a complex. Mol Biol Cell 2001; 12:2219-2228.
52. Lanzetti L, Rybin V, Malabarba MG et al. The Eps8 protein coordinates EGF receptor signalling through Rac and trafficking through Rab5. Nature 2000; 408:374-377.
53. Haas AK, Fuchs E, Kopajtich R et al. A GTPase-activating protein controls Rab5 function in endocytic trafficking. Nat Cell Biol 2005; 7:887-893.
54. Christoforidis S, Miaczynska M, Ashman K et al. Phosphoinositide-3-Kinases are Rab5 effectors. Nat Cell Biol 1999; 1:249-252.
55. Murray JT, Panaretou C, Stenmark H et al. Role of Rab5 in the recruitment of hVps34/p150 to the early endosome. Traffic 2002; 3:416-427.
56. Panaretou C, Domin J, Cockcroft S et al. Characterization of p150, an adaptor protein for the human phosphatidylinositol (PtdIns) 3-kinase. Substrate presentation by phosphatidylinositol transfer protein to the p150.Ptdins 3-kinase complex. J Biol Chem 1997; 272:2477-2485.
57. Simonsen A, Lippé R, Christoforidis S et al. EEA1 links phosphatidylinositol 3-kinase function to Rab5 regulation of endosome fusion. Nature 1998; 394:494-498.
58. Nielsen E, Christoforidis S, Uttenweiler-Joseph S et al. Rabenosyn-5, a novel Rab5 effector, is complexed with hVPS45 and recruited to endosomes through a FYVE finger domain. J Cell Biol 2000; 151:601-612.
59. Schnatwinkel C, Christoforidis S, Lindsay MR et al. The Rab5 effector Rabankyrin-5 regulates and coordinates different endocytic mechanisms. PLoS Biol 2004; 2:E261.
60. Stenmark H, Aasland R, Toh BH et al. Endosomal localization of the autoantigen EEA1 is mediated by a zinc- binding FYVE finger. J Biol Chem 1996; 271:24048-24054.
61. Pelkmans L, Burli T, Zerial M et al. Caveolin-stabilized membrane domains as multifunctional transport and sorting devices in endocytic membrane traffic. Cell 2004; 118:767-780.

62. Hoepfner S, Severin F, Cabezas A et al. Modulation of receptor recycling and degradation by the endosomal kinesin KIF16B. Cell 2005; 121:437-450.
63. Vanhaesebroeck B, Leevers SJ, Ahmadi K et al. Synthesis and function of 3-Phosphorylated inositol lipids. Annu Rev Biochem 2001; 70:535-602.
64. Shin HW, Hayashi M, Christoforidis S et al. An enzymatic cascade of Rab5 effectors regulates phosphoinositide turnover in the endocytic pathway. J Cell Biol 2005; 170:607-618.
65. Spaargaren M, Bos JL. Rab5 induces Rac-independent lamellipodia formation and cell migration. Mol Biol Cell 1999; 10:3239-3250.
66. Alvarez-Dominguez C, Barbieri AM, Beron W et al. Phagocytosed live Listeria monocytogenes influences Rab5-regulated in vitro phagosome-endosome fusion. J Biol Chem 1996; 271:13834-13843.
67. Perskvist N, Roberg K, Kulyte A et al. Rab5a GTPase regulates fusion between pathogen-containing phagosomes and cytoplasmic organelles in human neutrophils. J Cell Sci 2002; 115:1321-1330.
68. Fratti RA, Backer JM, Gruenberg J et al. Role of phosphatidylinositol 3-kinase and Rab5 effectors in phagosomal biogenesis and mycobacterial phagosome maturation arrest. J Cell Biol 2001; 154:631-644.
69. Barbieri MA, Roberts RL, Gumusboga A et al. Epidermal growth factor and membrane trafficking. EGF receptor activation of endocytosis requires Rab5a. J Cell Biol 2000; 151:539-550.
70. Barbieri MA, Fernandez-Pol S, Hunker C et al. Role of rab5 in EGF receptor-mediated signal transduction. Eur J Cell Biol 2004; 83:305-314.
71. Miaczynska M, Christoforidis S, Giner A et al. APPL proteins link Rab5 to nuclear signal transduction via an endosomal compartment. Cell 2004; 116:445-456.
72. Ikonomov OC, Sbrissa D, Shisheva A. Mammalian cell morphology and endocytic membrane homeostasis require enzymatically active phosphoinositide 5-kinase pikfyve. J Biol Chem 2001; 276:26141-26147.
73. Ikonomov OC, Sbrissa D, Mlak K et al. Active PIKfyve associates with and promotes the membrane attachment of the late endosome-to-trans-Golgi network transport factor Rab9 effector p40. J Biol Chem 2003; 278:50863-50871.
74. de Graaf P, Zwart WT, van Dijken RA et al. Phosphatidylinositol 4-kinasebeta is critical for functional association of rab11 with the Golgi complex. Mol Biol Cell 2004; 15:2038-2047.
75. Sciorra VA, Audhya A, Parsons AB et al. Synthetic genetic array analysis of the PtdIns 4-kinase Pik1p identifies components in a Golgi-specific Ypt31/rab-GTPase signaling pathway. Mol Biol Cell 2005; 16:776-793.
76. Rodriguez-Viciana P, Warne PH, Dhand R et al. Phosphatidylinositol-3-OH kinase as a direct target of Ras. Nature 1994; 370:527-532.
77. Snyder JT, Singer AU, Wing MR et al. The pleckstrin homology domain of phospholipase C-beta2 as an effector site for Rac. J Biol Chem 2003; 278:21099-21104.
78. Honda A, Nogami M, Yokozeki T et al. Phosphatidylinositol 4-phosphate 5-kinase alpha is a downstream effector of the small G protein ARF6 in membrane ruffle formation. Cell 1999; 99:521-532.
79. Jones DH, Morris JB, Morgan CP et al. Type I phosphatidylinositol 4-phosphate 5-kinase directly interacts with ADP-ribosylation factor 1 and is responsible for phosphatidylinositol 4,5-bisphosphate synthesis in the golgi compartment. J Biol Chem 2000; 275:13962-13966.
80. Christoforidis S, McBride HM, Burgoyne RD et al. The Rab5 effector EEA1 is a core component of endosome docking. Nature 1999; 397:621-625.
81. Christoforidis S, Zerial M. An affinity chromatography approach leading to the purification and identification of novel Rab effectors. Methods 2000; 20:403-410.
82. Gorvel J-P, Chavrier P, Zerial M et al. Rab5 controls early endosome fusion in vitro. Cell 1991; 64:915-925.
83. Rubino M, Miaczynska M, Lippe R et al. Selective membrane recruitment of EEA1 suggests a role in directional transport of clathrin-coated vesicles to early endosomes. J Biol Chem 2000; 275:3745-3748.
84. Nielsen E, Severin F, Backer JM et al. Rab5 regulates motility of early endosomes on microtubules. Nat Cell Biol 1999; 1:376-382.
85. Tsukazaki T, Chiang TA, Davison AF et al. SARA, a FYVE domain protein that recruits Smad2 to the TGFbeta receptor. Cell 1998; 95:779-791.
86. Itoh F, Divecha N, Brocks L et al. The FYVE domain in Smad anchor for receptor activation (SARA) is sufficient for localization of SARA in early endosomes and regulates TGF-beta/Smad signalling. Genes Cells 2002; 7:321-331.
87. Hayes S, Chawla A, Corvera S. TGF beta receptor internalization into EEA1-enriched early endosomes: Role in signaling to Smad2. J Cell Biol 2002; 158:1239-1249.
88. Hu Y, Chuang JZ, Xu K et al. SARA, a FYVE domain protein, affects Rab5-mediated endocytosis. J Cell Sci 2002; 115:4755-4763.

89. Le Roy C, Wrana JL. Clathrin- and nonclathrin-mediated endocytic regulation of cell signalling. Nat Rev Mol Cell Biol 2005; 6:112-126.
90. Hoeller D, Volarevic S, Dikic I. Compartmentalization of growth factor receptor signalling. Curr Opin Cell Biol 2005; 17:107-111.
91. Miaczynska M, Pelkmans L, Zerial M. Not just a sink: Endosomes in control of signal transduction. Curr Op Cell Biol 2004; 16:400-406.
92. Lanzetti L, Palamidessi A, Areces L et al. Rab5 is a signalling GTPase involved in actin remodelling by receptor tyrosine kinases. Nature 2004; 429:309-314.
93. Vitale G, Rybin V, Christoforidis S et al. Distinct Rab-binding domains mediate the interaction of Rabaptin-5 with GTP-bound Rab4 and Rab5. EMBO J 1998; 17:1941-1951.
94. de Renzis S, Sonnichsen B, Zerial M. Divalent Rab effectors regulate the sub-compartmental organization and sorting of early endosomes. Nat Cell Biol 2002; 4:124-133.
95. Fouraux MA, Deneka M, Ivan V et al. Rabip4' is an effector of rab5 and rab4 and regulates transport through early endosomes. Mol Biol Cell 2004; 15:611-624.
96. Lindsay AJ, Hendrick AG, Cantalupo G et al. Rab coupling protein (RCP), a novel Rab4 and Rab11 effector protein. J Biol Chem 2002; 277:12190-12199.
97. Stein MP, Feng Y, Cooper KL et al. Human VPS34 and p150 are Rab7 interacting partners. Traffic 2003; 4:754-771.
98. Ortiz D, Medkova M, Walch-Solimena C et al. Ypt32 recruits the Sec4p guanine nucleotide exchange factor, Sec2p, to secretory vesicles; evidence for a Rab cascade in yeast. J Cell Biol 2002; 157:1005-1015.
99. Walch-Solimena C, Collins RN, Novick P. Sec2 mediates nucleotide exchange on Sec4 and is involved in polarized delivery of post-Golgi vesicles. J Cell Biol 1997; 137:1495-1509.
100. Jedd G, Mulholland J, Segev N. Two new Ypt GTPases are required for exit from the yeast trans-Golgi compartment. J Cell Biol 1997; 137:563-580.
101. Benli M, Doring F, Robinson DG et al. Two GTPase isoforms, Ypt31p and Ypt32p, are essential for Golgi function in yeast. EMBO J 1996; 15:6460-6475.
102. Salminen A, Novick PJ. A ras-like protein is required for a post-Golgi event in yeast secretion. Cell 1987; 49:527-538.
103. Vonderheit A, Helenius A. Rab7 associates with early endosomes to mediate sorting and transport of Semliki forest virus to late endosomes. PLoS Biol 2005; 3:e233.
104. Rink J, Ghigo E, Kalaidzidis Y et al. Rab conversion as a mechanism of progression from early to late endosomes. Cell 2005; 122:735-749.
105. Hickson GR, Matheson J, Riggs B et al. Arfophilins are dual Arf/Rab 11 binding proteins that regulate recycling endosome distribution and are related to Drosophila nuclear fallout. Mol Biol Cell 2003; 14:2908-2920.
106. Novick P, Guo W. Ras family therapy: Rab, Rho and Ral talk to the exocyst. Trends Cell Biol 2002; 12:247-249.
107. Prigent M, Dubois T, Raposo G et al. ARF6 controls post-endocytic recycling through its downstream exocyst complex effector. J Cell Biol 2003; 163:1111-1121.
108. Zhang XM, Ellis S, Sriratana A et al. Sec15 is an effector for the Rab11 GTPase in mammalian cells. J Biol Chem 2004; 279:43027-43034.

CHAPTER 4

Synaptic Endosomes

Oleg Shupliakov*[1] and Volker Haucke[1]

Abstract

E ndosomes are important functional elements of the chemical synapse. They are used in membrane trafficking pathways controlling recycling and degradation of pre- and post-synaptic membrane proteins. Recent data indicate that they play a role in maintaining the pool of small synaptic vesicles and are involved in recycling of dense-core vesicle membrane during neurotransmitter release.

Membrane Trafficking Events at Synapses

Membrane trafficking in nerve cells appears to be more complex than in most other cell types. In addition to pathways common for nonneuronal cells, these cells utilize membrane trafficking mechanisms to release neuroactive substances into the surrounding environment.[1,2] These events occur to a large extent in specialized intracellular contacts established by neurons on target cells. These junctions are referred to as chemical synapses.

Chemical synapses are specialized signaling units composed of a pre- and a post-synaptic element. The postsynaptic element contains neurotransmitter receptors and protein machineries involved in signaling and receptor trafficking (see below). The presynaptic nerve terminal, in addition, contains neurotransmitter-filled organelles (vesicles), which may fuse with the presynaptic membrane. Neurons can secrete a variety of nonpeptidergic/classical and peptidergic transmitters via at least two types of secretory organelles, the small synaptic vesicles (SSVs) and the dense-core vesicles (DCVs), also referred to as secretory granules (Figs. 1 and 2A, B). According to the current model, the classical neurotransmitters acetylcholine (ACh), noradrenaline (NA), glutamate, glycine, and GABA are released from SSVs.[2] Neuropeptides, on the other hand, are stored in, and released from, DCVs,[3] which are directly formed at the trans-Golgi network and transported down the axon to their release sites.

Exocytosis of SSVs and DCVs is differentially regulated and takes place at different release sites of the nerve terminal.[4,5] SSVs empty their content upon depolarization and fusion of synaptic vesicles at defined regions of the presynaptic membrane. These areas contain a high density of calcium channels and protein complexes involved in vesicle docking and fusion and are referred to as "active zones".[2] DCVs tend to fuse outside the active zone region. Following neuroexocytosis, neurotransmitter molecules bind to postsynaptic receptors leading to an electrical response of the postsynaptic neuron. In addition, neurotransmitter release may lead to an activation of presynaptic receptors, which control retrograde modulation of neurotransmitter release. Recent data clearly demonstrate that receptors located on pre- and

*Corresponding Author: Oleg Shupliakov—Center of Excellence in Developmental Biology, Department of Neuroscience, Karolinska Institutet, 171 77 Stockholm, Sweden, Email: oleg.shupliakov@neuro.ki.s
[1]These authors contributed equally to this work.

Endosomes, edited by Ivan Dikic. ©2006 Landes Bioscience and Springer Science+Business Media.

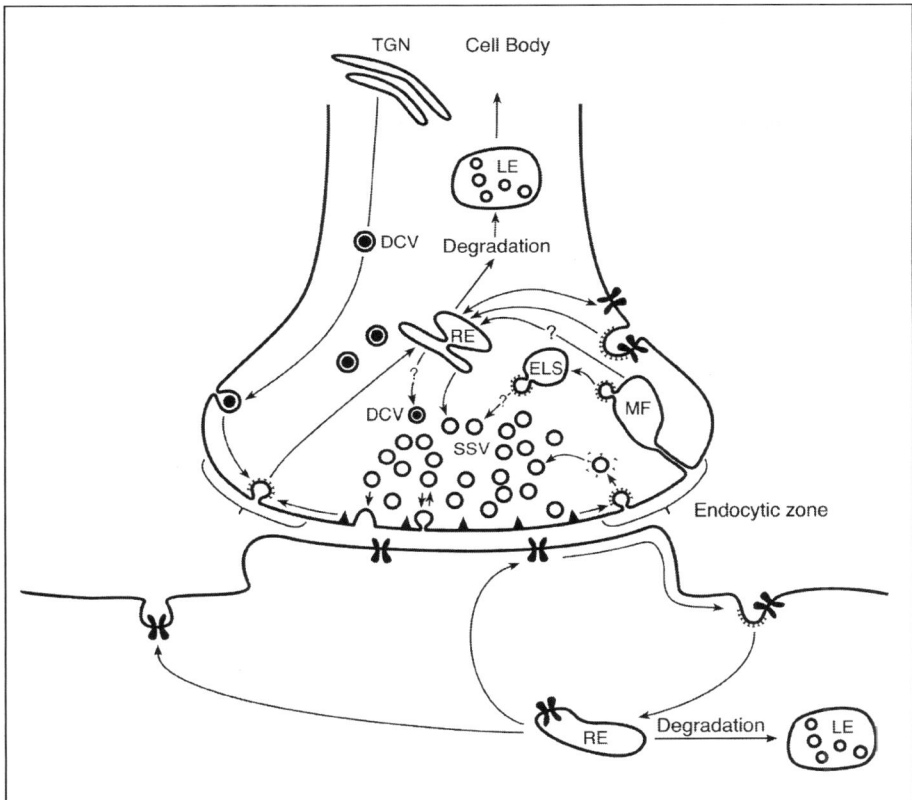

Figure 1. Membrane-trafficking pathways in synapses. Small synaptic vesicles (SSV) releasing neurotransmitter at the active zone (triangles) may be retrieved via "kiss and run" as well as via a clathrin-mediated mechanism. During intense activity, deep plasma membrane folds (MF) and endosome-like structures (ELS) are generated. Dense-cored vesicles (DCV) are synthesized at the TGN and are transported from the cell body to the synapse. They fuse predominantly outside the active zone. Their membrane may be retrieved by clathrin-dependent mechanisms. The 'classical' recycling pathway at synapses may regulate internalization and surface expression of receptors, transporters, and ion channels within both pre and post-synaptic compartments. This pathway involves early recycling endosomes (RE) as well as late endosomal compartments (LE, also referred as to multivesicular bodies), if a receptor undergoes degradation. Links in the pathways marked (?) remain to be elucidated. Depending on the physiological conditions, postsynaptic receptors may be retained in REs or targeted to late endosomes (LE) for degradation resulting in activity-dependent long-term depression (LTD). Under conditions favoring long-term potentiation (LTP) receptors can be exocytosed in an activity-dependent manner from recycling endosomal pools (RE) to extrasynaptic sites at the cell surface from where they are shuttled laterally to the synapse.

post-synaptic membranes can be retrieved from, or exposed at the membrane surface via distinct membrane-trafficking mechanisms, which may underlie synaptic plasticity phenomena.[6]

Several membrane retrieval mechanisms may function in a synaptic terminal. These involve uptake of membrane components related to SSVs, DCVs (see below), as well as receptors and ion channels. The latter mechanism resembles internalization of nutrient and signaling receptors in nonneuronal cells described in other chapters of this book.

Endocytic Recycling of Presynaptic Vesicles

Three mechanisms for synaptic vesicle endocytosis have been proposed: direct reformation of vesicles via the rapid closure of a transient fusion pore ("kiss-and-run"), clathrin-mediated endocytosis,[7] and bulk endocytosis.[8,9] During "kiss-and-run", SSVs are hypothesized to make brief contact with the plasma membrane forming a transient fusion pore through which the neurotransmitter is released.[2,10,11] In contrast, clathrin-mediated endocytosis occurs after complete fusion of SSVs with the plasma membrane.[2,12,13] The key components of the clathrin-dependent endocytosis machinery are: clathrin, the heterotetrameric adaptor complex (AP-2), and dynamin.[2,14] These proteins are part of the coat complex from very early stages. Recruitment of AP-2 to the plasma membrane is a complex process, which involves interactions with phosphoinositides, synaptotagmin,[10,15,16] and accessory proteins.[16] Although synapses use basically the same clathrin-dependent endocytic mechanism as nonneuronal cells, they utilize protein isoforms, most highly expressed in neurons. These include for example: AP180, auxillin, intersectins, dynamin-I, adaptin and the splice-variants of clathrin light chains.[2,14,17,18] AP180/CALM, epsins, intersectin and HIP1/HIP1R (huntingtin interacting proteins) function as cargo adaptors in addition to AP-2 (see chapter 10). While the clathrin lattice is formed, endophilin, epsin, and amphiphysin are involved in membrane invagination and clathrin rearrangements.[2,14,18] The GTPase dynamin is required for fisson of endocytic membrane vesicles.[17] Observation of clathrin-coated pit dynamics using total internal reflection microscopy indicates that during fission, dynamin recruitment to coated pits is rapidly followed by recruitment of actin.[19] Moreover perturbation of actin disrupts the endocytic reaction with accumulation of coated pits with wide necks[20] suggesting a role of actin and dynamin-interacting accessory proteins in promoting constriction of the neck. In lamprey, snake, and fly neuromuscular synapses, the invagination of the membrane into pits occurs at distinct "endocytic zones" surrounding the active zones of exocytosis. Distinct "hot-spots of endocytosis" have been also described at the post-synaptic membrane (Figs. 1 and 5; see also refs. 12,21).

Deep plasma membrane expansions and endosome-like compartments have been observed in synaptic terminals close to active zones during high-frequency stimulation of neuromuscular junctions, retina and also in central synapses.[7-9,22] They could be clearly seen in the lamprey giant synapse. The active zone in this junction is surrounded by organelle-free axoplasmic matrix. This allows following of these structures in serial ultrathin sections using electron microscopy (Fig. 2 C-I). Nonspecific membrane internalization by bulk endocytosis may prevent expansion of the cell surface under conditions in which clathrin-mediated endocytosis becomes rate limiting. Clathrin-coated buds are present not only at the plasma membrane but also on such endosome-like invaginations in synapses, consistent with the existence of the parallel pathway for clathrin-dependent synaptic vesicle formation (Figs. 1 and 2).

Presynaptic Endosomal Compartments

In analogy to nonneuronal cells endocytic vesicles presumably fuse with an endosomal compartment after detachment from the plasma membrane. Several studies performed recently support the involvement of bona fide endosomes in synaptic membrane recycling, although their role in different pathways still remains a matter of debate.[2,23] In hippocampal synapses for example, the role of endosomes in SSV recycling until recently had been believed to be limited. Studies using the fluorescent membrane dye FM1-43 have demonstrated that the amount of dye per vesicle taken up by endocytosis equals the amount of dye a vesicle releases upon exocytosis. It was thus concluded that the internalized vesicles participating in endo-exo recycling do not communicate with intermediate endosomal compartments during the recycling process.[24] These experiments, however, did not exclude the possibility that a remaining, "second" population of vesicles, not participating in exo-endocytic recycling, could exchange membrane with an endosome or that this organelle could be recruited upon certain activity demands. Several recent studies have provided evidence that the population of SSVs is

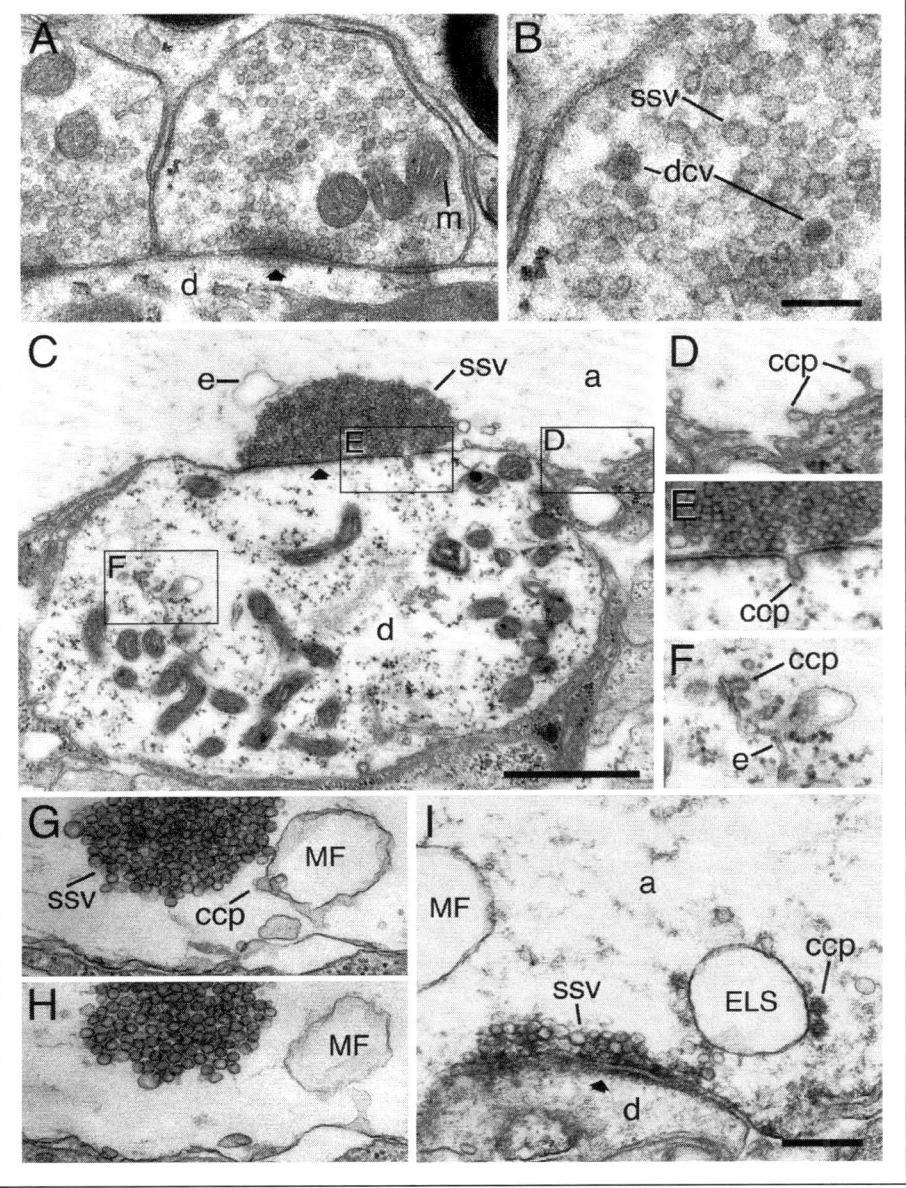

Figure 2. Ultrastructure of endocytic intermediates. A) Electron micrograph of a chemical synapse from the cat spinal cord. An area of the terminal containing small synaptic vesicles (SSV) and dense-core vesicles (DCV) is shown in (B) at higher magnification. C) A reticulospinal synapse in the lamprey spinal cord fixed during recovery after high-frequency stimulation. Boxed areas show clathrin-mediated intermediates budding from the presynaptic endocytic zone (D), from the post-synaptic membrane (E), and from an endocytic compartment of the post-synaptic dendrite, d (F)je-putative endosome. G and H) Serial ultrathin sections from the endocytic zone of a lamprey giant synapse stimulated with high K$^+$ (30 mM) showing a plasma membrane fold

(MF) containing a clathrin-coated intermediate (ccp). (I) Electron micrograph of a giant synapse stimulated with high K⁺. Note the endosome-like compartment (ELS; as revealed from serial sections), containing elathrin-coated intermediates on its surface. Designations: a-axoplasmic matrix; m- mitochondrion. Thick arrows indicate active zones. Bars: A, D-I - 300 nm; B - 200 nm; C - 500 nm. (Shupliakov, unpublished observations; see also Gad et al., 1998).

indeed inhomogenous, consistent with the idea that different pools of vesicles may use distinct membrane-trafficking pathways during the synaptic vesicle cycle. It has been shown, that during development, vesicular release along growing axons of frog motoneurons in culture is sensitive to brefeldin A (BFA), whereas quantal release from nerve terminals is BFA-insensitive.[25] It cannot be excluded that a similar mechanism may be retained in adult synapses. Studies in hippocampal synapses, for example, show that spontaneously recycling vesicles and activity-dependently recycling vesicles originate from distinct pools with limited crosstalk with each other.[26]

Early Endosomal Compartments

Recent studies at the *Drosophila* neuromuscular junction have provided direct support for the involvement of an endosomal pathway in the synaptic vesicle cycle.[27] For a number of years it has been known that the small GTPase Rab5 is present on isolated SSVs.[23,27-29] By recruitment of several effector molecules Rab5 promotes the formation of endosomes in nonneuronal cells.[30,31] Active Rab5 recruits two phosphatidylinositol-3-kinases, PI(3)-kinases p85/p110 and VPS34/p150, which trigger a local enrichment of phosphatidylinositol- 3-phosphate, PI(3)P, in the endosomal membrane.[32] PI(3)P specifically binds to the FYVE zinc-finger domain of endosomal factors such as the Rab5 effectors EEA1 and Rabenosyn-5, which ultimately mediate endocytic vesicle tethering and fusion with early endosomes.[33-36] Consistently, blocking of PI(3)-kinases with antibodies or wortmannin impairs the association of FYVE domain proteins with early endosomes thereby, blocking endosomal membrane trafficking.[35,37] FYVE domains binds to PI(3)P within an intact lipid bilayer[38,39] and the localization of a myc-tagged tandem repeat of the FYVE domain (myc-2xFYVE) is restricted to early endosomes and the internal membrane of multivesicular bodies.[40] Thus, both Rab5 and 2xFYVE can be considered as selective markers for PI(3)P-containing endosomes. Using these two GFP-tagged markers as well as antibodies, González-Gaitán and colleagues have recently demonstrated the presence of Rab5-positive, PI(3)P-containing endosomes at the presynaptic terminal of *Drosophila* neuromuscular junctions.[27] Under conditions in which the SSV pool was depleted, endosomes were drastically reduced in size and recovered by dynamin-mediated endocytosis. Interfering with Rab5 function using a dominant-negative version of Rab5 caused a reduction in the number of released quanta during synaptic transmission, whereas elevated levels of Rab5 increased the quantal content. These data indicate that Rab5-dependent trafficking pathway plays a role in presynaptic vesicle cycling.

Support for the involvement of an early endosomal compartment in synaptic membrane trafficking also comes from studies on the endosomal membrane adaptor complex AP-3. AP-3, which exists as both ubiquitously expressed AP-3A as well as a neuron-specific AP-3B isoform, is localized to the TGN and/or endosomal compartments. It participates in trafficking to the vacuole/lysosome in yeast,[41,42] flies,[43-45] and mammals.[46,47] AP-3B as well as ADP ribosylation factor 1 (ARF1)[48,49] are required for the biogenesis of synaptic-like microvesicles budding from PC12 cell endosomes.[49-52] Genetic analysis of AP-3 mutant mice has been linked to a variety of neurological defects.[50,53-55] The *mocha* mouse, a null mutation of the δ subunit of AP-3, exhibits balance and hearing problems, is hyperactive, and is prone to seizures.[50,53-55] Mice in which the neuron-specific AP-3B subunit μ3B has been genetically deleted, show specific defects related to the biogenesis of GABA-containing SSVs suggesting a particularly important function for AP-3B at inhibitory synapses.[56]

Another potential function for endosome-derived synaptic vesicles and AP-3 dependent pathway is in the recovery of membrane components of dense core vesicles (DCVs) that have

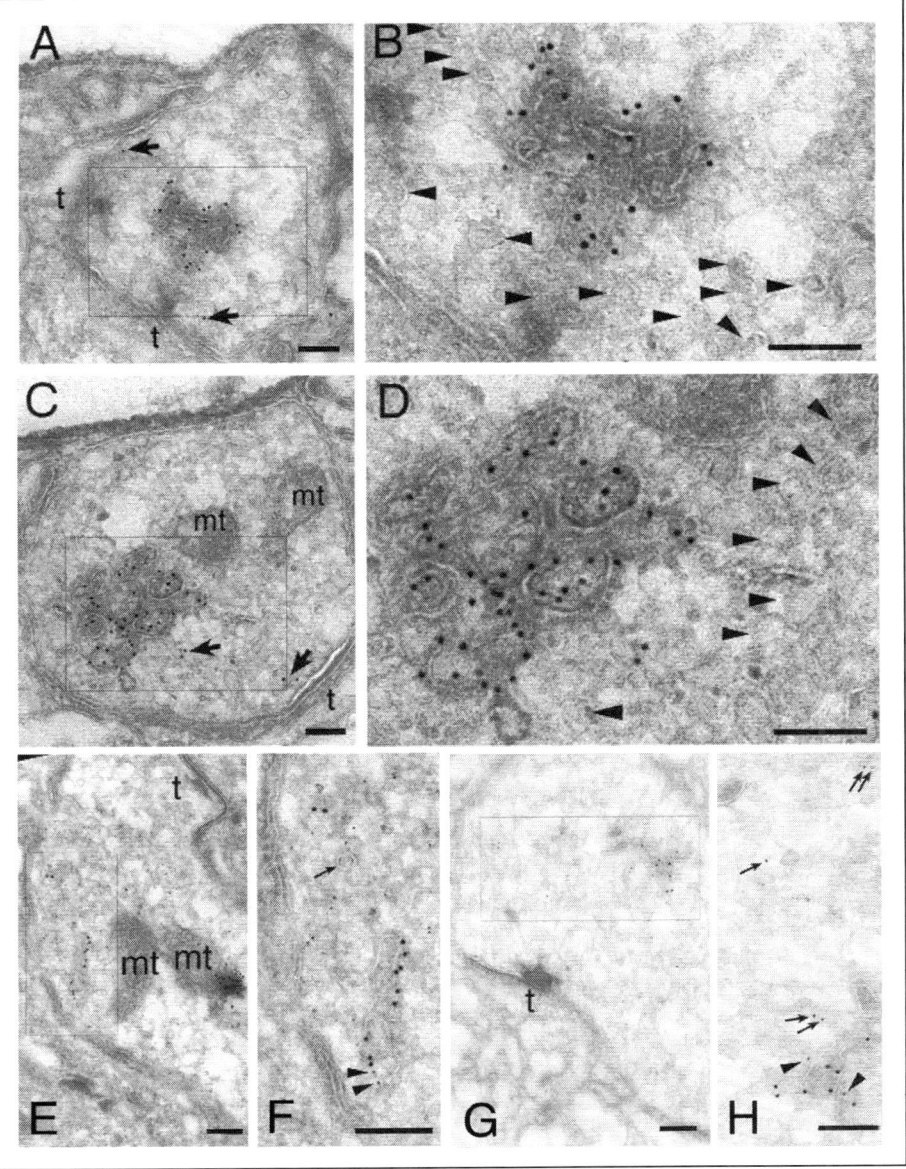

Figure 4. Cryoimmuno-EM of the GFP-2xFYVE endosome. A-D) Cryoimmuno-electron micrographs showing two *Drosophila* presynaptic terminals (A/B and C/D), where GFP-2xFYVE is labeled by 10-nm gold particles (anti-GFP antibody). B and D) High magnifications of the boxes in A and C, respectively. We found cisternal structures of around 150 nm associated to a more electron-dense region within the terminal. The darker regions allow a better contrast for visualization of the membrane (which appear lighter in cryosections) associated to the endosomes, compared with the vesicles with a diameter of 35 or 70 nm. Vesicles are, however, occasionally observed (arrowheads). Only few gold particles (7.8±1.3%, *n*= 5 sections) are associated with the vesicles (arrows). Cryoimmuno-electron micrographs showing localization of GFP-2xFYVE

(10-nm gold particles) and endogenous CSP (5 nm gold; E and F), or endogenous Rab5 (5 nm gold; G and H). F and H) High magnifications of the boxes in E and G, respectively. E and F) CSP appears throughout the bouton area associated with the pool of vesicles, whereas GFP-2xFYVE is largely restricted to the cisternal endosomal compartments. Although not many vesicles are distinguishable (F, arrow), their presence is revealed by staining of SV integral membrane protein CSP. Few 5-nm gold particles labeling CSP could also be observed in the cisternal structures (F, arrowheads). Rab5 appears in the cisternal structures, (H, arrowheads) as well as in other regions corresponding to vesicles or cytosol (H, arrows). t, T-bar or electron-dense regions indicating active zones; mt, mitochondria. Bars: (A-D) 150 nm; (E-H) 200 nm *(reproduced with permission from Wucherpfennig et al., 2004).*

just undergone exocytosis. Membrane retrieval of this type has been followed in PC12 cells transfected with a chimeric P-selectin.[52,57,58] It had been proposed that, neuronal AP-3B may recapture protein components of DCV proteins. A recapture step could sequester selected DCV proteins from a degradative pathway and allow them to be incorporated into the synaptic vesicle cycle. The distribution of neuronal AP-3B showed some resemblance to that reported for chromogranin A, a marker of dense core granules, particularly in the stratum oriens and the molecular layer of the dentate gyrus.[59] These data indirectly support a role for AP-3B in the recovery of DCV-derived membrane components.

Late Endosomal Compartments

A more acidic late endosomal compartment has been shown to form during maturation of early endosomes in nonneuronal cells.[1] Whereas early endosomes tend to be tubular and are located towards the cell periphery, late endosomes are more spherical and often appear closer to the nucleus. A subset of late endosomes has a multivesicular appearance, hence named multivesicular bodies (MVBs). Late endosomes form a dynamic network together with lysosomal structures, the end point of endocytosis and site of protein degradation. As mentioned above, transport of DCV membrane constituents in neurons involves early endosomes,[58] whereas multivesicular bodies may particpate in retrograde transport of DCV components towards the cell body. Such transport has been observed i.e., in the splenic nerve.[60]

Endosomes in Postsynaptic Receptor Trafficking

Over the past few years it has become clear that the strength of synaptic connections, in particular with respect to postsynaptic responses, is subject to plastic changes. At excitatory synapses, activation of glutamate receptors, such as AMPA-type glutamate-gated ion channels provides the primary depolarization in excitatory neurotransmission. AMPA receptor-mediated postsynaptic currents are modulated by changes in their localization and surface expression. Glutamate receptor density thus appears to be carefully regulated by fine-tuning receptor synthesis, endosomal trafficking, and degradation.[6] Since most of what we know about endosomal trafficking of postsynaptic receptors has been derived from studies on excitatory glutamate receptors we will focus primarily on these, but it is expected that similar mechanisms are utilized for other receptor types as well. In agreement with this notion it has been reported that ionotropic GABA$_A$[61-63] and glycine[64,65] receptors regulating inhibitory neurotransmission in the nervous system, can also be internalized into endosomal or subsynaptic compartments.

AMPA Receptors Are Internalized via Constitutive or Ligand-Induced Pathways

AMPA receptors, heterotetramers composed of related GluR1-4 subunits, undergo dynamic redistribution in and out of the postsynaptic membrane. Most excitatory synapses form on dendritic spines, that emanate from the main shaft and usually bear a single synaptic contact at their heads. AMPA receptors, concentrated at the postsynaptic density (PSD) of dendritic spines, serve to propagate the signal[66] and are able to dynamically move into and out of the postsynaptic density by lateral diffusion. They may also undergo constitutive internalization.[67]

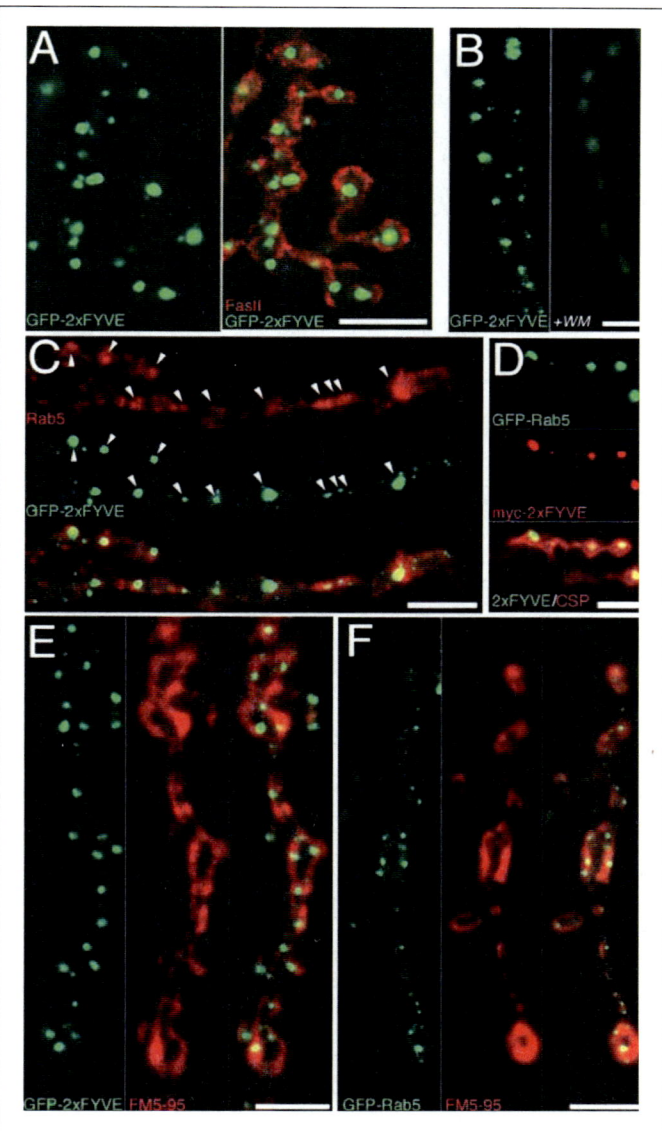

Figure 3. An endosomal compartment at the presynaptic terminal. A) Double labeling showing GFP-2xFYVE (green) to monitor the endosomes and Fasciclin II immunostaining to label the NMJ presynaptic terminals (FasII, red). B) GFP-2xFYVE fluorescence in an abdominal muscles 6/7 NMJ before (left) and after (right) a 45min treatment with 100 nM wortmannin in vivo. Note that, upon wortmannin treatment, GFP-2xFYVE loses the punctate pattern and becomes dispersed into the cytosol. Untreated controls retained the punctate pattern. C) Double labeling showing endogenous Rab5 immunostaining (red) and GFP-2xFYVE (green); lower panel shows merge. Arrowheads indicate Rab5 punctate structures colocalizing with GFP-2xFYVE-positive endosomes. Notice also that some of the Rab5 endosomes do not contain GFP-2xFYVE, consistent with two

types of Rab5 endosomes, EEA1 positive/negative. D) Triple labeling showing GFP-Rab5 (top, green), endosomal myc-2xFYVE immunostaining using an anti-c-myc antibody (middle, red; bottom, green) and CSP immunostaining (CSP, bottom, red) to label the presynaptic terminals in a muscles 6/7 NMJ. Bottom panel is a merge of myc-2xFYVE and CSP. GFP-Rab5 and myc-2xFYVE show a complete colocalization. Double labelings in green (E) GFP-2xFYVE or (F) GFP-Rab5 and in red FM5-95 styryl dye internalized into the presynaptic terminal upon a 1-min stimulation with 60 mM K⁺ to label the pool of recycling vesicles (E and F) in two different abdominal muscles 6/7 NMJs. Right panels show merge. Note that the endosomes are embedded within the pool of recycling vesicles. NMJs from late third instar larvae. Genotypes: (A-C and E) *w; UAS-GFP-myc-2xFYVE; elav-GAL4*; D) *w; UASmyc- 2xFYVE/elav-GAL4 UAS-GFP-Rab5*; and (F) *w; elav-GAL4/UASGFP-Rab5*. Bars, 5 μm. *(reproduced with permission from Wucherpfennig et al., 2004).*

Stimulation of glutamatergic synapses with AMPA, NMDA, or insulin has been shown to enhance AMPA receptor internalization by clathrin-mediated endocytosis.[67-71] AMPA receptor internalization along the endocytic pathway correlates physiologically with activity-dependent long-term depression (LTD). Conversely, during long-term potentiation (LTP), a cellular model for learning and memory, an increase in the number of functional, cell-surface exposed AMPA receptors at the postsynaptic membrane is observed (Fig. 1; see also refs. 6,72). These receptors are thought to originate from an intracellular reserve pool.[73,74] Endocytic removal of AMPA receptors occurs mostly from extrasynaptic sites.[75] This observation is consistent with the predominant localization of endocytic proteins including clathrin, AP-2, and dynamin lateral to the postsynaptic density.[76]

The exact molecular mechanisms of the constitutive and regulated pathways for AMPA receptor internalization are not yet completely understood. Although all pathways are dependent on the GTPase dynamin, an accessory protein required for fission of both clathrin- and nonclathrin-coated vesicles, and its SH3 domain-containing binding partners,[67] they seem to be spatially segregated and differentially influenced by protein kinases,[77] phosphatases, and calcium ions.[68,70]

AMPA Receptors Undergo Differential Endosomal Sorting

Different stimuli differentially affect the subcellular localization of internalized receptors. AMPA receptors endocytosed via direct agonist stimulation (i.e., AMPA) colocalize with early endosomal markers such as early endosomal antigen 1 (EEA1), syntaxin 13, and endocytosed transferrin receptors. In contrast, AMPA receptors internalized via insulin- or NMDA-regulated signaling pathways although initially present in EEA1-positive early endosomes appear to segregate into distinct compartments, which may include late endosomes and lysosomes;[78] but see[77] for a different view). How precisely and at which stage differential endosomal sorting occurs remains unclear. Activated AMPA receptors colocalize with AP-2[69] and Eps15[67] in clathrin-coated pits. Direct binding of the basic stretch within the cytoplasmic tail of the AMPA receptor, subunits GluR1-3, to the clathrin adaptor complex AP-2 is only required, for NMDA-induced AMPA receptor endocytosis,[79] thus indicating that differential recognition modes at the cell surface may contribute to endosomal sorting. In nematodes, GluR is subject to multi-ubiquitination, which may target glutamate receptors for internalization and late endosomal/ lysosomal degradation.[80] Differential sorting of receptors recognized directly by endocytic adaptors or modified by ubiquitination is seen in nonneuronal cells, i.e., in the case of internalized transferrin vs. epidermal growth factor receptors.[81,82] Additionally, insulin-stimulated AMPA receptor internalization may be regulated by tyrosine phosphorylation,[83] similar to what is seen for growth factor receptors.[81]

Receptor Determinants for Endosomal Sorting

As discussed above AMPA receptors internalized in response to direct agonist binding (i.e., AMPA) or NMDA-induced signaling cascades initially share the same early endosomal sorting

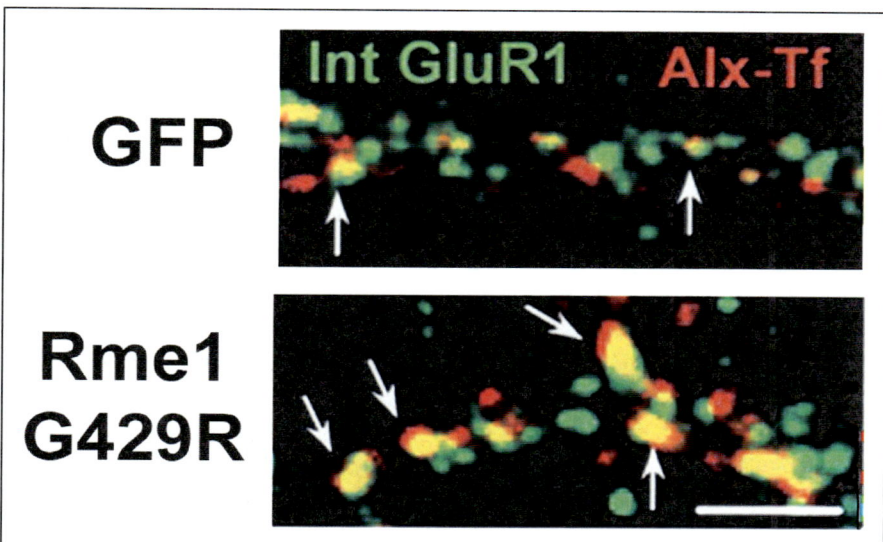

Figure 5. Recycling endosomes supply AMPA receptors for long-term potentiation (LTP) Overexpressing a mutant version of the Eps15-homology domain protein EHD1/ Rme1 (Rme1-G429R) traps internalized AMPA receptor GluR1 (green) in recycling endosomes where it colocalizes (yellow puncta, see arrows) with endocytosed Alexa-labeled transferrin (red). Scale bar, 2 μm. Under such conditions LTP is abolished. *(reproduced with permission from Park et al., 2004).*

pathway.[78,84] During AMPA-induced internalization, homomeric GluR2 receptors appear to be retained within early recycling endosomes, whereas GluR2 endocytosed in response to NMDA is diverted to late endosomes and lysosomes for degradation.[78] One important factor regulating sorting appears to be the subunit composition of heteromeric AMPA receptors. Homomeric GluR1 receptors are retained in recycling endosomes, whereas GluR3 homomers enter the late endosomal/ lysosomal pathway regardless of stimulation.[78] In the context of heteromers endosomal sorting is apparently governed by GluR2, which exerts dominant effects, perhaps by recruiting adaptor proteins,[85] by undergoing posttranslational modifications including tyrosine phosphorylation[83] and ubiquitination[80] or by binding to ubiquitinated adaptor proteins such as PSD-95.[86] In nonneuronal cells, sorting of ubiquitinated cargo is achieved by ubiquitin-interacting motif (UIM) or ubiquitin-associated domain (UBA) containing accessory proteins including the phosphoinositide binding protein epsin, the EH-domain containing endocytic accessory protein Eps15, and Hrs.[81,82] Both epsin and Eps15 are highly expressed in the brain and could serve functions in postsynaptic receptor sorting within the endosomal system, similar to their known roles in presynaptic vesicle recycling. In the case of the inhibitory GABA$_A$ receptor channel, it has recently been demonstrated that huntingtin-associated protein 1 (HAP1) modulates cell surface receptor number by inhibiting lysosomal GABA$_A$ receptor degradation.[87] Since HAP1 can associate with the ubiquitin-binding adaptor Hrs[85] it is tempting to speculate that HAP1 may act by suppressing Hrs-dependent lysosomal receptor targeting. Although HAP1 action appears to be restricted to inhibitory synapses similar regulatory principles may hold true for early endosomal trafficking of glutamate receptors.

Whereas endosomal targeting of AMPA receptors during conditions of long-term depression (LTD) is well established, much less is known about the recycling of internalized receptors to the cell surface. Recent data suggest that indeed recycling endosomes rather than trans-Golgi network (TGN)-derived vesicles supply AMPA receptors for long-term potentiation (LTP). Blocking exit from recycling endosomes by expression of dominant-negative mutants of either Rab11a or the EH-domain containing accessory protein EHD1/ Rme1 trapped internalized AMPA receptors in recycling endosomes (Fig. 5) and prevented expression of LTP in hippocampal slices.[84]

Thus, early recycling endosomes appear to play crucial roles in synaptic plasticity by regulating the internalization, recycling, degradation, and thus cell surface number of glutamate and possibly other ionotropic receptors at synapses.

References

1. Le Roy C, Wrana JL. Clathrin- and nonclathrin-mediated endocytic regulation of cell signalling. Nat Rev Mol Cell Biol 2005; 6(2):112-126.
2. Murthy VN, De Camilli P. Cell biology of the presynaptic terminal. Annu Rev Neurosci 2003; 26:701-728.
3. Torrealba F, Carrasco MA. A review on electron microscopy and neurotransmitter systems. Brain Res Brain Res Rev 2004; 47(1-3):5-17.
4. Bruns D, Jahn R. Real-time measurement of transmitter release from single synaptic vesicles. Nature 1995; 377(6544):62-65.
5. Lundberg JM, FrancoCereceda A, Lou YP et al. Differential release of classical transmitters and peptides. Adv Second Messenger Phosphoprotein Res 1994; 29:223-234.
6. Bredt DS, Nicoll RA. AMPA receptor trafficking at excitatory synapses. Neuron 2003; 40(2):361-379.
7. Heuser JE, Reese TS. Evidence for recycling of synaptic vesicle membrane during transmitter release at the frog neuromuscular junction. J Cell Biol 1973; 57(2):315-344.
8. Gad H, Low P, Zotova E et al. Dissociation between Ca2+-triggered synaptic vesicle exocytosis and clathrin-mediated endocytosis at a central synapse. Neuron 1998; 21(3):607-616.
9. Takei K, Mundigl O, Daniell L et al. The synaptic vesicle cycle: A single vesicle budding step involving clathrin and dynamin. J Cell Biol 1996; 133(6):1237-1250.
10. Galli T, Haucke V. Cycling of synaptic vesicles: How far? How fast! Sci STKE 2004; 2004(264):re19.
11. Valtorta F, Meldolesi J, Fesce R. Synaptic vesicles: Is kissing a matter of competence? Trends Cell Biol 2001; 11(8):324-328.
12. Jarousse N, Kelly RB. Endocytotic mechanisms in synapses. Curr Opin Cell Biol 2001; 13(4):461-469.
13. Wenk MR, De Camilli P. Protein-lipid interactions and phosphoinositide metabolism in membrane traffic: Insights from vesicle recycling in nerve terminals. Proc Natl Acad Sci USA 2004; 101(22):8262-8269.
14. Szymkiewicz I, Shupliakov O, Dikic I. Cargo- and compartment-selective endocytic scaffold proteins. Biochem J 2004; 383(Pt 1):1-11.
15. Craxton M. Synaptotagmin gene content of the sequenced genomes. BMC Genomics 2004; 5(1):43.
16. Takei K, Haucke V. Clathrin-mediated endocytosis: Membrane factors pull the trigger. Trends Cell Biol 2001; 11(9):385-391.
17. Hinshaw JE. Dynamin and its role in membrane fission. Annu Rev Cell Dev Biol 2000; 16:483-519.
18. McMahon HT, Mills IG. COP and clathrin-coated vesicle budding: Different pathways, common approaches. Curr Opin Cell Biol 2004; 16(4):379-391.
19. Merrifield CJ. Seeing is believing: Imaging actin dynamics at single sites of endocytosis. Trends Cell Biol 2004; 14(7):352-358.
20. Shupliakov O, Bloom O, Gustafsson JS et al. Impaired recycling of synaptic vesicles after acute perturbation of the presynaptic actin cytoskeleton. Proc Natl Acad Sci USA 2002; 99(22):14476-14481.
21. Blanpied TA, Scott DB, Ehlers MD. Dynamics and regulation of clathrin coats at specialized endocytic zones of dendrites and spines. Neuron 2002; 36(3):435-449.
22. Paillart C, Li J, Matthews G et al. Endocytosis and vesicle recycling at a ribbon synapse. J Neurosci 2003; 23(10):4092-4099.
23. de Hoop MJ, Huber LA, Stenmark H et al. The involvement of the small GTP-binding protein Rab5a in neuronal endocytosis. Neuron 1994; 13(1):11-22.
24. Murthy VN, Stevens CF. Synaptic vesicles retain their identity through the endocytic cycle. Nature 1998; 392(6675):497-501.
25. Zakharenko S, Chang S, O'Donoghue M et al. Neurotransmitter secretion along growing nerve processes: Comparison with synaptic vesicle exocytosis. J Cell Biol 1999; 144(3):507-518.

26. Sara Y, Virmani T, Deak F et al. An isolated pool of vesicles recycles at rest and drives spontaneous neurotransmission. Neuron 2005; 45(4):563-573.
27. Wucherpfennig T, Wilsch-Brauninger M, Gonzalez-Gaitan M. Role of Drosophila Rab5 during endosomal trafficking at the synapse and evoked neurotransmitter release. J Cell Biol 2003; 161(3):609-624.
28. Fischer von Mollard G, Stahl B, Walch-Solimena C et al. Localization of Rab5 to synaptic vesicles identifies endosomal intermediate in synaptic vesicle recycling pathway. Eur J Cell Biol 1994; 65(2):319-326.
29. Shimizu H, Kawamura S, Ozaki K. An essential role of Rab5 in uniformity of synaptic vesicle size. J Cell Sci 2003; 116(Pt 17):3583-3590.
30. de Renzis S, Sonnichsen B, Zerial M. Divalent Rab effectors regulate the sub-compartmental organization and sorting of early endosomes. Nat Cell Biol 2002; 4(2):124-133.
31. Sonnichsen B, De Renzis S, Nielsen E et al. Distinct membrane domains on endosomes in the recycling pathway visualized by multicolor imaging of Rab4, Rab5, and Rab11. J Cell Biol 2000; 149(4):901-914.
32. Christoforidis S, McBride HM, Burgoyne RD et al. The Rab5 effector EEA1 is a core component of endosome docking. Nature 1999; 397(6720):621-625.
33. Lawe DC, Patki V, Heller-Harrison R et al. The FYVE domain of early endosome antigen 1 is required for both phosphatidylinositol 3-phosphate and Rab5 binding. Critical role of this dual interaction for endosomal localization. J Biol Chem 2000; 275(5):3699-3705.
34. Nielsen E, Christoforidis S, Uttenweiler-Joseph S et al. Rabenosyn-5, a novel Rab5 effector, is complexed with hVPS45 and recruited to endosomes through a FYVE finger domain. J Cell Biol 2000; 151(3):601-612.
35. Simonsen A, Lippe R, Christoforidis S et al. EEA1 links PI(3)K function to Rab5 regulation of endosome fusion. Nature 1998; 394(6692):494-498.
36. Stenmark H, Vitale G, Ullrich O et al. Rabaptin-5 is a direct effector of the small GTPase Rab5 in endocytic membrane fusion. Cell 1995; 83(3):423-432.
37. Mills IG, Jones AT, Clague MJ. Involvement of the endosomal autoantigen EEA1 in homotypic fusion of early endosomes. Curr Biol 1998; 8(15):881-884.
38. Misra S, Hurley JH. Crystal structure of a phosphatidylinositol 3-phosphate-specific membrane-targeting motif, the FYVE domain of Vps27p. Cell 1999; 97(5):657-666.
39. Sankaran VG, Klein DE, Sachdeva MM et al. High-affinity binding of a FYVE domain to phosphatidylinositol 3-phosphate requires intact phospholipid but not FYVE domain oligomerization. Biochemistry 2001; 40(29):8581-8587.
40. Gillooly DJ, Morrow IC, Lindsay M et al. Localization of phosphatidylinositol 3-phosphate in yeast and mammalian cells. EMBO J 2000; 19(17):4577-4588.
41. Cowles CR, Odorizzi G, Payne GS et al. The AP-3 adaptor complex is essential for cargo-selective transport to the yeast vacuole. Cell 1997; 91(1):109-118.
42. Stepp JD, Huang K, Lemmon SK. The yeast adaptor protein complex, AP-3, is essential for the efficient delivery of alkaline phosphatase by the alternate pathway to the vacuole. J Cell Biol 1997; 139(7):1761-1774.
43. Kretzschmar D, Poeck B, Roth H et al. Defective pigment granule biogenesis and aberrant behavior caused by mutations in the Drosophila AP-3beta adaptin gene ruby. Genetics 2000; 155(1):213-223.
44. Mullins C, Hartnell LM, Wassarman DA et al. Defective expression of the mu3 subunit of the AP-3 adaptor complex in the Drosophila pigmentation mutant carmine. Mol Gen Genet 1999; 262(3):401-412.
45. Ooi CE, Moreira JE, Dell'Angelica EC et al. Altered expression of a novel adaptin leads to defective pigment granule biogenesis in the Drosophila eye color mutant garnet. EMBO J 1997; 16(15):4508-4518.
46. Le Borgne R, Alconada A, Bauer U et al. The mammalian AP-3 adaptor-like complex mediates the intracellular transport of lysosomal membrane glycoproteins. J Biol Chem 1998; 273(45):29451-29461.
47. Yang W, Li C, Ward DM et al. Defective organellar membrane protein trafficking in Ap3b1-deficient cells. J Cell Sci 2000; 113(Pt 22):4077-4086.
48. Faundez V, Horng JT, Kelly RB. ADP ribosylation factor 1 is required for synaptic vesicle budding in PC12 cells. J Cell Biol 1997; 138(3):505-515.
49. Horng JT, Tan CY. Biochemical characterization of the coating mechanism of the endosomal donor compartment of synaptic vesicles. Neurochem Res 2004; 29(7):1411-1416.
50. Blumstein J, Faundez V, Nakatsu F et al. The neuronal form of adaptor protein-3 is required for synaptic vesicle formation from endosomes. J Neurosci 2001; 21(20):8034-8042.
51. Faundez V, Horng JT, Kelly RB. A function for the AP3 coat complex in synaptic vesicle formation from endosomes. Cell 1998; 93(3):423-432.
52. Hannah MJ, Schmidt AA, Huttner WB. Synaptic vesicle biogenesis. Annu Rev Cell Dev Biol 1999; 15:733-798.

53. Kantheti P, Qiao X, Diaz ME et al. Mutation in AP-3 delta in the mocha mouse links endosomal transport to storage deficiency in platelets, melanosomes, and synaptic vesicles. Neuron 1998; 21(1):111-122.
54. Miller CL, Burmeister M, Stevens KE. Hippocampal auditory gating in the hyperactive mocha mouse. Neurosci Lett 1999; 276(1):57-60.
55. Vogt K, Mellor J, Tong G et al. The actions of synaptically released zinc at hippocampal mossy fiber synapses. Neuron 2000; 26(1):187-196.
56. Nakatsu F, Okada M, Mori F et al. Defective function of GABA-containing synaptic vesicles in mice lacking the AP-3B clathrin adaptor. J Cell Biol 2004; 167(2):293-302.
57. Blagoveshchenskaya AD, Norcott JP, Cutler DF. Lysosomal targeting of P-selectin is mediated by a novel sequence within its cytoplasmic tail. J Biol Chem 1998; 273(5):2729-2737.
58. Partoens P, Slembrouck D, Quatacker J et al. Retrieved constituents of large dense-cored vesicles and synaptic vesicles intermix in stimulation-induced early endosomes of noradrenergic neurons. J Cell Sci 1998; 111(Pt 6):681-689.
59. Munoz DG. The distribution of chromogranin A-like immunoreactivity in the human hippocampus coincides with the pattern of resistance to epilepsy-induced neuronal damage. Ann Neurol 1990; 27(3):266-275.
60. Annaert WG, Quatacker J, Llona I et al. Differences in the distribution of cytochrome b561 and synaptophysin in dog splenic nerve: A biochemical and immunocytochemical study. J Neurochem 1994; 62(1):265-274.
61. Kittler JT, Delmas P, Jovanovic JN et al. Constitutive endocytosis of GABAA receptors by an association with the adaptin AP2 complex modulates inhibitory synaptic currents in hippocampal neurons. J Neurosci 2000; 20(21):7972-7977.
62. van Rijnsoever C, Sidler C, Fritschy JM. Internalized GABA-receptor subunits are transferred to an intracellular pool associated with the postsynaptic density. Eur J Neurosci 2005; 21(2):327-338.
63. Herring D, Huang R, Singh M et al. Constitutive GABAA receptor endocytosis is dynamin-mediated and dependent on a dileucine AP2 adaptin-binding motif within the beta 2 subunit of the receptor. J Biol Chem 2003; 278(26):24046-24052.
64. Buttner C, Sadtler S, Leyendecker A et al. Ubiquitination precedes internalization and proteolytic cleavage of plasma membrane-bound glycine receptors. J Biol Chem 2001; 276(46):42978-42985.
65. Rasmussen H, Rasmussen T, Triller A et al. Strychnine-blocked glycine receptor is removed from synapses by a shift in insertion/degradation equilibrium. Mol Cell Neurosci 2002; 19(2):201-215.
66. Sheng M, Lee SH. AMPA receptor trafficking and the control of synaptic transmission. Cell 2001; 105(7):825-828.
67. Man HY, Lin JW, Ju WH et al. Regulation of AMPA receptor-mediated synaptic transmission by clathrin-dependent receptor internalization. Neuron 2000; 25(3):649-662.
68. Beattie EC, Carroll RC, Yu X et al. Regulation of AMPA receptor endocytosis by a signaling mechanism shared with LTD. Nat Neurosci 2000; 3(12):1291-1300.
69. Carroll RC, Lissin DV, von Zastrow M et al. Rapid redistribution of glutamate receptors contributes to long-term depression in hippocampal cultures. Nat Neurosci 1999; 2(5):454-460.
70. Lin JW, Ju W, Foster K et al. Distinct molecular mechanisms and divergent endocytotic pathways of AMPA receptor internalization. Nat Neurosci 2000; 3(12):1282-1290.
71. Luscher C, Xia H, Beattie EC et al. Role of AMPA receptor cycling in synaptic transmission and plasticity. Neuron 1999; 24(3):649-658.
72. Malinow R, Malenka RC. AMPA receptor trafficking and synaptic plasticity. Annu Rev Neurosci 2002; 25:103-126.
73. Passafaro M, Sheng M. Synaptogenesis: The MAP location of GABA receptors. Curr Biol 1999; 9(7):R261-263.
74. Pickard L, Noel J, Duckworth JK et al. Transient synaptic activation of NMDA receptors leads to the insertion of native AMPA receptors at hippocampal neuronal plasma membranes. Neuropharmacology 2001; 41(6):700-713.
75. Ashby MC, De La Rue SA, Ralph GS et al. Removal of AMPA receptors (AMPARs) from synapses is preceded by transient endocytosis of extrasynaptic AMPARs. J Neurosci 2004; 24(22):5172-5176.
76. Racz B, Blanpied TA, Ehlers MD et al. Lateral organization of endocytic machinery in dendritic spines. Nat Neurosci 2004; 7(9):917-918.
77. Ehlers MD. Reinsertion or degradation of AMPA receptors determined by activity-dependent endocytic sorting. Neuron 2000; 28(2):511-525.
78. Lee SH, Simonetta A, Sheng M. Subunit rules governing the sorting of internalized AMPA receptors in hippocampal neurons. Neuron 2004; 43(2):221-236.

79. Lee SH, Liu L, Wang YT et al. Clathrin adaptor AP2 and NSF interact with overlapping sites of GluR2 and play distinct roles in AMPA receptor trafficking and hippocampal LTD. Neuron 2002; 36(4):661-674.
80. Burbea M, Dreier L, Dittman JS et al. Ubiquitin and AP180 regulate the abundance of GLR-1 glutamate receptors at postsynaptic elements in C. elegans. Neuron 2002; 35(1):107-120.
81. Dikic I. Mechanisms controlling EGF receptor endocytosis and degradation. Biochem Soc Trans 2003; 31(Pt 6):1178-1181.
82. Hicke L, Dunn R. Regulation of membrane protein transport by ubiquitin and ubiquitin-binding proteins. Annu Rev Cell Dev Biol 2003; 19:141-172.
83. Ahmadian G, Ju W, Liu L et al. Tyrosine phosphorylation of GluR2 is required for insulin-stimulated AMPA receptor endocytosis and LTD. EMBO J 2004; 23(5):1040-1050.
84. Park M, Penick EC, Edwards JG et al. Recycling endosomes supply AMPA receptors for LTP. Science 2004; 305(5692):1972-1975.
85. Li Y, Chin LS, Levey AI et al. Huntingtin-associated protein 1 interacts with hepatocyte growth factor-regulated tyrosine kinase substrate and functions in endosomal trafficking. J Biol Chem 2002; 277(31):28212-28221.
86. Colledge M, Snyder EM, Crozier RA et al. Ubiquitination regulates PSD-95 degradation and AMPA receptor surface expression. Neuron 2003; 40(3):595-607.
87. Kittler JT, Thomas P, Tretter V et al. Huntingtin-associated protein 1 regulates inhibitory synaptic transmission by modulating gamma-aminobutyric acid type A receptor membrane trafficking. Proc Natl Acad Sci USA 2004; 101(34):12736-12741.

CHAPTER 5

Endosome Fusion

Dorothea Brandhorst and Reinhard Jahn*

Abstract

In recent years it has become apparent that membrane fusion reactions in the secretory pathway are mediated by supramolecular assemblies that include both members of conserved protein families and proteins specific for individual fusion steps. Before fusion, membranes need to recognize and bind to each other, and it is thought that this step is mediated by Rab-GTPases and their effectors. Fusion itself is probably mediated by SNARE proteins but other factors such as SM-proteins are also involved. In this chapter, we discuss fusion reactions of the endocytotic pathway, with particular emphasis on homotypic fusion between mammalian early and late endosomes, and between yeast vacuoles, with a main emphasis on the molecular mechanism of SNARE proteins.

Introduction: Fusion Steps in the Endocytotic Pathway

The endocytic pathway in higher eukaryotic cells comprises pleiotropic intracellular organelles enclosed by single membranes. These organelles are connected with each other by vesicular traffic that includes distinct budding, transport, and fusion steps (Fig. 1, see also Chapter 1). By definition, the starting point of the endocytic pathway is endocytosis, i.e., the formation of invaginations and the pinching off of transport vesicles from the plasma membrane, which include both clathrin-coated and noncoated vesicles, specialized organelles such as phagosomes, and caveolae. With exception of caveolae and phagosomes, the first compartment reached is the early/sorting endosome with which endocytotic vesicles fuse.[1,2] Early endosomes must be considered as an intracellular distribution center from which trafficking pathways lead back to the plasma membrane and to late endosomes/lysosomes. In addition they are connected to the Golgi membrane system. Early endosomes are in dynamic equilibrium and rapidly fuse not only with incoming endocytotic vesicles but also with each other ("homotypic" fusion).[3-5] Steady-state is maintained by the parallel and continuous generation of transport vesicles with different destinations. A first pathway leads directly back to the plasma membrane and mediates recycling of certain receptors. A second pathway involves the formation of cisternal vesicles that are transported to the perinuclear region. They form a separate compartment, termed recycling endosomes, from where membranes and membrane resident proteins are also returned to the plasma membrane, albeit at a slower rate than by the direct pathway. Recycling endosomes also communicate directly with the trans Golgi Network. Third, early endosomes ship vesicles back to the TGN[6] and in turn receive TGN-derived vesicles.[7] Finally, trafficking to late endosomes is thought to occur by maturation of early endosomes that develop into multivesicular bodies/late endosomes.[8,9] Alternatively, it cannot be excluded that early endosomes are more stable compartments that give rise to transport intermediates that are then capable of fusing with late endosomes.[10,11]

*Corresponding Author: Reinhard Jahn—Department of Neurobiology, Max-Planck-Institute for Biophyiscal Chemistry, Am Fassberg, 37077 Göttingen, Germany. Email: rjahn@gwdg.de

Endosomes, edited by Ivan Dikic. ©2006 Landes Bioscience and Springer Science+Business Media.

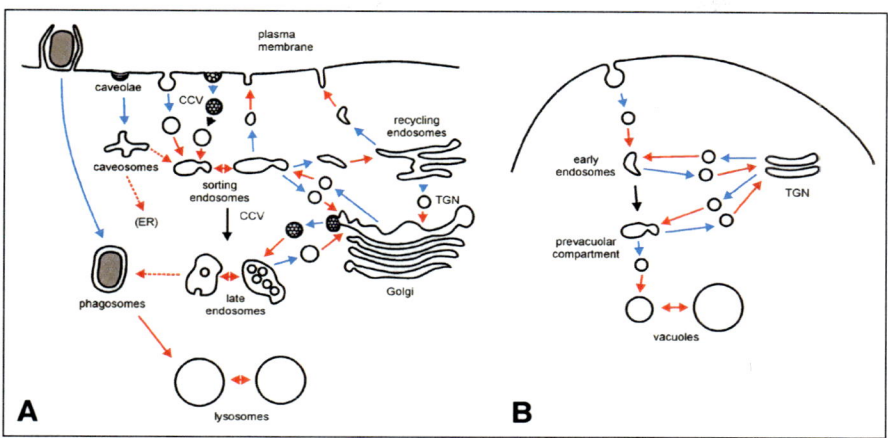

Figure 1. Major trafficking routes of the endocytotic pathway in a mammalian cell (left) and yeast (right). Blue arrows, budding/fission followed by transport; red arrows, transport followed by fusion; black arrows, transport. CCV, clathrin-coated vesicle; TGN, trans-Golgi network. Note that the autophatocytotic pathway is not included.

Like early endosomes, late endosomes are capable of homotypic fusion. Furthermore, they fuse with lysosomes,[12-14] and finally, lysosomes fuse homotypically with each other.[15,16] Late endosomes are also connected by vesicular traffic to the Golgi system.[17,18] It needs to be borne in mind that the list of fusion reactions indicated in Figure 1 may still be incomplete. For instance, in polarized cells apical sorting endosomes do not fuse with basolateral endosomes, suggesting further differentiation of trafficking steps that may also occur in nonpolarized cells.[19] Furthermore, it is becoming apparent that certain cisternal compartments (e.g., early endosomes) are further differentiated into subdomains with distinct functions,[20,21] and it is thus conceivable that a single organelle possesses specialized sites for distinct fusion reactions.

Endocytotic trafficking is also studied intensely in yeast. Although due to the small size of yeast cells imaging approaches are more limited and challenging, the powerful combination of genetics and convenient in vitro fusion reactions led to the discovery of a string of novel proteins, many of which, as it turns out, are conserved from yeast to mammals.[22] Genetics is facilitated by the fact that yeast survives (albeit often poorly) when endocytotic trafficking steps are impaired or completely dysfunctional. In general, the yeast endocytic pathway is similar to that of mammalian cells. It comprises endocytic vesicles, early endosomes, late endosomes that are also termed prevacuolar compartment, and the vacuolar compartment which is considered as equivalent to mammalian lysosomes (Fig. 1B, for review see ref. 23). Fusion reactions take place between endocytic vesicles and early endosomes, between early endosomes and the prevacuolar compartment, and between the prevacuolar compartment and vacuoles. Furthermore, vacuoles fragment during cell division and then coalesce by homotypic fusion. The latter fusion reaction has been intensely studied using both genetic and in vitro approaches.[24-26] Finally, trafficking routes exist between the trans Golgi network and the endosomal membrane system (Fig. 1B).[27,28]

In recent years, we have witnessed an explosive growth in our understanding of intracellular fusion reactions. Conserved protein families have been identified that are responsible for distinct steps in the fusion pathway. While our understanding of some model fusion reactions is already quite advanced, others are much less understood. For instance, it is not known to which extent the fusion reactions indicated in Figure 1 are different at the molecular level. While it is clear that homotypic fusion between early and late endosomes, respectively, involves different sets of proteins, fusion of endocytotic vesicles with early endosomes appears to share the fusion machinery with homotypic fusion of early endosomes.[29] Another example includes the homotypic fusions of late endosomes and lysosomes, respectively, that appear to involve similar proteins.[30] In these cases, it is not known whether all proteins are shared, and if so, how these fusions are distinguished.

For a successful fusion reaction the membranes need to recognize the appropriate fusion partner, get in close physical contact, and finally fuse their lipid bilayers. The key players in these intracellular fusion reactions are represented by conserved protein families including the small, ras-related Rab/Ypt GTPases, the SM proteins, and the SNAREs. These proteins are interacting with a plethora of additional proteins, many of which are apparently for specific fusion reactions and whose function is often only incompletely understood. In this chapter, we summarize recent developments in the fusion reactions of the endocytotic pathway, with the main emphasis on the fusion reaction itself.

Tethering and Docking

Before fusion, the corresponding organelles need to be brought into close proximity, resulting in physical contact. Little is known about these initial reactions. There are some hints that physical contact between organelles might be sufficient to trigger a rection cascade leading to fusion.[31] Studying tethering and docking requires the availability of in vitro assays, and consequently most of the information is derived from the homotypic fusion of early endosomes in mammalian cells, and from the homotypic fusion of yeast vacuoles. In early endosome fusion, candidate proteins were identified that are likely to be involved in tethering: EEA1 (early endosomal antigen 1) forms long coiled coils that have the capacity to bind to endosomes with two binding sites on both ends and act as long-range tethering factor.[32] EEA1 was found to bind phosphatidylinositol 3 phosphate (PI(3)P) on early endosomes via a FYVE domain and to Rab5. Rab5 belongs to the family of small GTPases that cycle between GTP- and GDP-bound forms. Probably, GTP-Rab5, in conjunction with the local generation of PI(3)P mediates the initial assembly of docking complexes. In addition to EEA1, GTP-Rab5 recruits a plethora of effectors to early endosomes (including a PI-3-kinase) and thereby provides a platform for the assembly of the fusion machinery.[33] This process is discussed in detail in chapter 5 (Zerial).

Homotypic vacuole fusion in yeast in vitro proceeds in four consecutive and discernable steps: priming, tethering, docking and fusion. In the priming step, SNAREs are disassembled (see below, refs. 34,35), a prerequisite for the following tethering reaction. The tethering reaction again involves a Rab-protein (Ypt7p, the yeast orthologue to mammalian Rab7) as central player in the initial steps. Ypt7p is activated by a multiprotein complex referred to as HOPS (homotypic vacuole fusion and protein sorting) or Class C Vps complex that consists of six different proteins termed Vps11p, Vps16p, Vps 18p, Vps 33p (an SM protein, see section *SM-Proteins: Essential but Still Enigmatic*), Vps39p, and Vps 41p.[36-38] Ypt7p activation is needed for docking reactions that interestingly do not seem to involve long tethering proteins such as EEA1. Rather, an involvement of the actin cytoskeleton is suggested by the requirement for activating Rho-GTPases. While tethering by activation of Ypt7 is a reversible process, the subsequent formation of trans SNARE complexes (see below) leads to irreversibly docked vacuoles.[39] Microscopic studies of in vitro docked vacuoles showed that SNARE proteins, the HOPS complex, and Ypt7 are enriched at the rim of the contact sites.[40,41]

The two examples discussed above illustrate the central role of Rab/Ypt proteins in orchestrating docking and tethering. There are about 60 Rabs in mammals and 11 Ypts in yeast.[42,43] Each of them interacts with its own set of partially very diverse effector proteins. While some of these effectors may be shared between subsets of rabs, others are specific. Currently, the Rab/Ypt proteins, together with their effectors, their GTPase activating and guanine nucleotide exchange factors, are the best candidates for determining the specificity of intracellular fusion reactions. Biologically, this makes sense since "proofreading" of the membrane is most efficient if it occurs early in the reaction cascade instead of late, i.e., when docking machineries are already assembled.

SNARE-Proteins: Central Players in Membrane Fusion

General Properties of SNAREs

SNARE proteins (NSF attachment protein receptors[44] are small (10-35 kDa), mostly membrane bound proteins that each contain a homologous stretch of 60-70 amino acids, referred to as SNARE motif. SNARE proteins undergo an assembly-disassembly cycle. In solution, monomeric SNARE motifs are largely unstructured. When appropriate SNAREs are combined, they form a stable α-helical bundle, called SNARE complex.[45] The crystal structures of the neuronal and the late endosomal SNARE complexes have been solved and show a high degree of structural similarity. They consist of coiled coil bundles that contain four different α-helices.[46,47] The core of the helical bundle is built by 16 layers of highly conserved, mostly hydrophobic amino acid sidechains. Due to a highly conserved polar residue in the center of the SNARE motif which ist either arginine or glutamine, SNAREs are classified into Qa-SNAREs (syntaxins), Qb-SNAREs, Qc-SNAREs and R-SNAREs.[48] Although not conclusively proven in all cases, it is becoming clear that each functional SNARE complex contains one SNARE motif of each of these subfamilies ("QabcR-rule"). The four SNARE motifs can be provided by three different proteins as in the neuronal or by four proteins as in the late endosomal SNARE complex. SNARE complexes show an extraordinary stability towards heat and denaturants, for instance they "melt" only at temperatures above 75°C.[49]

How do SNARE complexes form? The four helices in the SNARE complex are aligned in parallel, with the transmembrane domains at one end and the N-termini at the other end. This led to the so called "zippering model".[50-53] The model proposes that assembly of the SNAREs starts at the N-termini of the SNARE motifs and proceeds towards the C-terminally localized membrane anchor domains. The energy provided by the assembly reaction could be high enough to overcome the repulsion forces of the negatively charged lipid headgroups in the membranes and finally fuse the lipid bilayers. During the fusion reaction, complexes shift from "trans" (transmembrane domains in two different lipid bilayers) to "cis" (transmembrane domains in the same lipid bilayer). According to this model, the function of the SNAREs is to execute membrane merger. Indeed, it has been shown that appropriate sets of SNARE proteins reconstituted into proteoliposomes fuse membranes, albeit at a slow rate.[53-56] However, as discussed further below (see section *Late Steps in the Fusion Pathway and Fusion Catalysis*), this view is not undisputed, and other steps downstream of the SNAREs have been invoked before bilayer fusion.

Disassembly of SNARE complexes is mediated by NSF (N-ethylmaleimide sensitive factor), a hexameric ATPase belonging to the family of AAA proteins (ATPases associated with other activities).[57,58] It requires cofactors called SNAPs (soluble NSF acceptor proteins) for this reaction. In mammals, three isoforms of SNAPs have been identified, α-, β- and γ-SNAP with α-SNAP being the ubiquitous isoform.[59] Three α-SNAPs bind to one SNARE complex, then NSF can bind, resulting in a so called 20S complex.[44] The exact mechanism of SNARE complex disassembly is still unclear. Activation of the NSF ATPase activity by α-SNAP is needed for the disassembly reaction as shown by a dominant negative α-SNAP mutant.[60] A single NSF possesses six identical subunits,[61] each containing a catalytically active ATPase site, thus providing enough energy to disassemble this extremely stable complex. The disassembly of SNARE complexes is essential for fusion, since SNARE proteins are bound in complexes after the fusion reaction and have to be dissociated.

For successful fusion, each membrane has to contain at least one SNARE with a transmembrane domain. In the first characterized SNARE complex involved in neuronal exocytosis, the R-SNARE synaptobrevin is localized to synaptic vesicles while the Q-SNAREs Syntaxin 1 (Qa) and SNAP 25 (Qbc) reside on the plasma membrane. Therefore, SNAREs were originally classified into v-SNARE (vesicular SNARE) and t-SNAREs (target membrane SNAREs).[44] In many heterotypic fusion events, R-SNAREs are found on vesicles and Q-SNAREs on the target membrane. However, there are exeptions such as the fusion reaction of transport vesicles from the ER with the cis-Golgi in yeast. Here, the vesicle contains R-, Qb- and Qc-SNARE while the Qa-SNARE is found on the target membrane. Furthermore, v- and t-SNAREs cannot be distinguished in homotypic fusion events. In addition, certain SNARE proteins were shown to participate in SNARE complexes with different topologies (see below), further complicating the classification into v- and t-SNAREs (see ref. 62 for a more comprehensive discussion). For these reasons, the structurally based classification into Q- and R-SNAREs is preferable.

SNAREs in the Endosomal Pathway

The original SNARE hypothesis proposed that each individual intracellular fusion reaction is catalysed by a unique set of SNARE proteins that thus would represent the major determinants for the specificity of intracellular fusion reactions.[55,57] However, subsequent work revealed that the situation is more complicated, for the following reasons:

- In vitro studies using purified proteins revealed that SNAREs promiscuously form complexes as long as the QabcR rule is followed although not all of these complexes are as stable as the cognate complexes.[63,64] In liposome fusion experiments, a higher degree of specificity was suggested but these experiments are not conclusive since in many cases the QabcR rule was not followed.[55,65] Promiscuity is in fact not surprising when considering the extraordinary high degree of structural conservation between evolutionary distant SNARE complexes.[46]

- While several lines of evidence indicate that each trafficking step requires specific sets of SNAREs, deletion of individual SNAREs in both yeast and mammals have shown that in some fusion steps certain SNAREs are at least partially redundant. For instance, knock-out mice deficient in the Qb-SNARE vti1b that was previously shown to function in late endosome fusion are viable, and despite pleiotropic phenotypes endocytotic trafficking to the lysosome does not appear to be grossly impaired.[66] Intriguingly, its partner syntaxin 8 (Qc) is downregulated in the absence of vti1b. So either other Qb/Qc-SNAREs are able to substitute, or a second SNARE complex operates in parallel in the same fusion reaction that hitherto has escaped detection. Another example is the R-SNARE Nyv1p that functions in yeast vacuole fusion. Again, deletion results in mild phenotypes,[34] which is probably due to substitution by the R-SNARE Ykt6p.[35]

- Individual SNAREs can participate in several fusion reactions, each involving different SNARE partners. Yeast Vti1p is used throughout the endosomal system as component of four different SNARE complexes, functioning in traffic from the Golgi to the endosome, from the Golgi to the vacuole, in retrotrade traffic to the cis-Golgi, and in homotypic fusion of the TGN.[67] Other examples include the R-SNARE Ykt6p that functions both in homotypic fusion of vacuoles, and in ER-Golgi transport,[35,68] and the R-SNARE Sec22p that functions both in anterograde and retrograde traffic between the endoplasmic reticulum and the cis-Golgi, each step involving different Q-SNARE partners.[69]

- A further complication arises from the fact that as integral membrane proteins, SNAREs need to follow membrane recycling pathways after completing their task in a fusion reaction in order to return to their prefusion compartment. For example, SNAREs involved in exocytosis in neuroendocrine cells recycle through early endosomes. Cleavage of these SNAREs with clostridial neurotoxins has no effect on homotypic fusion[70] suggesting that these SNAREs are not involved. Thus, regulatory factors are needed that distinguish which of the SNAREs to use in an upcoming fusion event and which to silence because they are merely travelling passengers. The nature of these factors is unknown.

Considering these complications it is not surprising that it has been difficult to unequivocally assign SNAREs and SNARE complexes to individual fusion steps of the endosomal

Table 1. SNAREs in endosomal fusion steps

Fusion Step	SNARE Candidates	References
Mammals		
EE-EE/CCV-EE	Qa syntaxin 13, syntaxin16	97,98
	Qb vti1a	
	Qc syntaxin 6	
	R VAMP-4/VAMP-8	
LE-LE	Qa syntaxin 7	46,72,97
	Qb vti1b	
	Qc syntaxin 8	
	R VAMP-8	
LE-Lys	Qa syntaxin 7	14
	Qb vti1b	
	Qc syntaxin 8	
	R VAMP-7	
EE/RE-TGN	Qa syntaxin16, syntaxin 5	99,100
	Qb vti1a, GS28	
	Qc syntaxin 6, GS15	
	R VAMP-4, VAMP-3, Ykt6	
Yeast		
Vac-Vac, Prevac-Vac	Qa Vam3p	35,101
	Qb vti1p	
	Qc Vam7p	
	R Nyv1p, Ykt6p	
TGN-Prevac	Qa Pep12p	102
	Qb Vti1p	
	Qc Syn8p	
	R Ykt6p	

EE: early endosomes; LE: late endosomes; CCV: clathrin-coated vesicles; RE: recycling endosomes; Lys: lysosomes; TGN: trans-Golgi network; Vac: vacuole; Prevac: prevacuolar compartment

pathway, particularly in mammalian cells. Unfortunately, there are no fast-acting tools available that inactivate endosomal SNAREs, unlike exocytotic SNAREs that are selectively cleaved by botulinum and tetanus neurotoxins.[71] Inhibition by antibodies is notoriously unreliable due to steric hindrance, and competition by excess soluble SNAREs is counteracted by the NSF-disassembly reaction, thus further limiting the experimental options for assigning SNAREs to individual reactions.

Table 1 shows the SNAREs that are presently discussed for the fusion steps of the endocytic pathway. It needs to be emphasized that in particular in mammalian cells the evidence for the involvement of a particular SNARE is frequently limited to coprecipitation or to perturbation of in vitro fusion using antibodies and recombinant proteins, and more evidence is required to affirm the involvement of given sets of SNAREs in a particular fusion step.

In mammalian cells, most authors agree that the SNAREs syntaxin 7 (Qa), syntaxin 8 (Qb), syntaxin 8 (Qc), and endobrevin/VAMP8 (R) catalyse the fusion of late endosomes, and a crystal structure for this complex is available.[46,72] However, as discussed above, endosomal trafficking is not blocked in mice lacking vti1b[66] or endobrevin/VAMP8,[73] strongly suggesting the involvement of other SNAREs. For the other fusion steps, the responsible SNARE complexes are even less clear. For instance, for the homotypic fusion of early endosomes the Qa-SNAREs syntaxin 13 and syntaxin16, the Qb-SNAREs vti1a, the Qc-SNARE syntaxin 6, and the R-SNAREs VAMP-4 and VAMP-8/endobrevin are discussed. Syntaxin 6 and syntaxin 13 were found to bind EEA1.[74,75] These findings suggest a link between SNAREs and the tethering complex orchestrated by Rab5, similar to the link between Ypt7p/HOPS and the SNAREs in vacuolar fusion. On the other hand, syntaxin 16 interacts with the SM protein Vps45 that appears to be specific for early endosome fusion, indicating a role for this Qa-SNARE in the same fusion step.[76] To complicate things further, SNAP-25 has recently been invoked as a Qb/c SNARE in the fusion of early endosomes.[77] SNAP-25 has a well-established role in regulated exocytosis of synaptic vesicles and secretory granules, where it interacts with syntaxin 1 (Qa) and synaptobrevin/VAMP2.[44,57]

In yeast, the evidence invoking specific sets of SNAREs to fusion steps in the endocytotic pathway is stronger. For instance, in homotypic fusion of vacuoles the SNAREs in charge of this reaction include Vam3p (Qa), Vam7p (Q), vti1 (Q), and Nyv1p (R). In addition, Ykt6p can substitute for Nyv1p. Similarly, the identity of the SNAREs involved in fusions at the prevacuolar compartments are better established than those in the correspoinding fusion steps of mammalian cells (Table 1).

SM-Proteins: Essential but Still Enigmatic

SM proteins (Sec1/munc18 like proteins) are hydrophilic proteins of 60-70 kDa that bind to SNARE proteins. Their arch-shaped overall structure seems to be conserved (for review see refs. 78,79). As there are fewer SM proteins than fusion reactions in the cell (7 in the human genome, 4 in yeast), it is assumed that a given SM protein acts in several fusion steps. Fusion is absolutely dependent on the corresponding SM proteins, but the details of their molecular role are not well understood. While most authors agree that SM proteins somehow regulate SNAREs, the mechanism of action is controversial. Furthermore, several SM proteins are—directly or indirectly—linked to Rab/Ypt proteins by means of forming complexes with Rab effectors.

The crystal structure of two only distantly related SM proteins (mammalian Munc18, and yeast Sly1p) show strong structural conservation.[80,81] Furthermore, most SM proteins bind to Qa-SNAREs but surprisingly the nature of the binding interface is not conserved: Syntaxin 1 binds to a large cleft in the SM-protein Munc-18, with syntaxin being folded up in a "closed" conformation.[80] In stark contrast, the Qa-SNARE Sed5p interacts only with few N-terminal amino acids with its SM-partner Sly1p, and binding does not occur to the cleft (that is conserved) but to the outer surface of the globular molecule.[81] In line with these differences, the effects of Qa-SNARESM interaction on SNARE pairing are divergent: Munc18 binding to syntaxin 1 prevents SNARE complex formation, Sly1p or Vps45 binding to the corresponding Qa-SNAREs appears to promote SNARE pairing, and in yet another case (Sec1) the SM protein binds only to the assembled SNARE complex.[78,79,82,83]

Two SM proteins have been identified that operate in fusion events of the endocytotic pathway both in yeast and mammals, including Vps45 and Vps33, and they nicely exemplify the problems in arriving at a common concept for these molecules. Vps45 binds to an N-terminal peptide of syntaxin 16/Tlg2p.[76,84] Syntaxin 16 is one of the Qa-SNARE candidates for an early endosomal SNARE complex (see above). In absence of Vps45p, Tlg2p is no longer capable of forming complexes with its SNARE partners.[85] The Rab5 effector protein Rabenosyn-5 was found to build a complex with Vps45, acting as a linker molecule between Rab and SM protein on early endosomes.[86] Vps33 is mainly studied in yeast where it functions in homotypic fusion of vacuoles. As mentioned above, unlike other SM proteins Vps33p is part of a multiprotein complex (HOPS or Vps/C) that interacts both with Rabs and SNAREs. Clearly, more work is needed before the still enigmatic role of these important proteins in membrane fusion is understood.

Late Steps in the Fusion Pathway and Fusion Catalysis

As discussed above, the SNAREs are the best candidates for fusion catalysts, and this view is supported by a large body of evidence. However, primarily studies on the homotypic fusion of yeast vacuoles (the "last" fusion step in the endocytotic pathway) provided evidence that additional reactions are required downstream of the SNAREs before fusion. Using primarily the effects of inhibitors (such as antibodies) on the kinetics of in vitro fusion of vacuoles as argument, SNAREs pairing was primarily assigned to the docking reaction, and deemed expendable for the subsequent steps leading to fusion.[87] Events that were shown by these approaches to occur downstream of SNAREs include the action of a protein phosphatase, the release of calcium from vacuolar stores, and finally the "trans"-complexation of a certain type of the Vo-subunit of the vacuolar proton ATPase that is thought to be activated by calmodulin and then form a proteinaceous fusion pore.[22,26,88,89]

Docking causes a release of calcium from the vacuole.[90] In fact, many intracellular fusion reactions are dependent on the local release of calcium (including homotypic fusion of mammalian early endosomes or of ER-membranes). These fusions are completely blocked by fast calcium chelators, which can be overcome by readding free calcium. More recently, it was reported that SNAREpairing in "trans" during vacuole docking is directly responsible for the Ca^{2+} release.[91]

The target of the released calcium in the fusion pathway is calmodulin that selectively binds to the vacuolar membrane in a Ca^{2+}-dependent manner. The Ca^{2+}-calmodulin receptor on the membrane was subsequently identified as the Vo domain of the vacuolar proton ATPase.[88] This enzyme is conserved from archebacteria to higher eukaryotes and is responsible for the acidification of the lumen of the entire intracellular endomembrane system. Structurally, it has many similarities with the mitochondrial FoF1-ATPase but it is not capable of synthesizing ATP. Recently, it has been shown that the V-ATPase also has a head and a stalk domain that rotate with respect to the membrane-embedded proteolipid ring in the membrane, made up of Vo subunits.[92] The discovery that calmodulin appears to mediate its activation of vacuolar fusion via binding to Vo, and that this activation occurred downstream of SNARE involvement, seriously challenged the view that SNAREs are fusion catalysts. Rather, an alternative model was proposed according to which two proteolipid rings, arranged in "trans", would form a connexon-like fusion pore that would enlarge during fusion.[93]

Presently, it cannot yet be decided which of the two mutually exclusive models for the fusion mechanism is correct, although admittedly we have a hard time understanding that the conserved Vo subunit of a highly conserved proton pump shall have a "second life" (also referred to as "moonlighting") as a fusion pore. Recently, however, it has been shown that the block exerted by calcium chelators on yeast vacuole fusion can be completely rescued by adding the recombinant SNARE Vam7p.[94] This SNARE that operates as Qc-SNARE in vacuole fusion belongs to a small subgroup of SNAREs that do not possess a membrane anchor. Rather, Vam7p cycles off and on the membrane,[95] probably being aided by its phox-homology (PX) domain that selectively binds to phosphoinositol-3-phosphate.[96] The fact that Vam7p bypasses the chelator block seriously challenges the notion that the calcium-calmodulin system operates downstream of SNAREs, thus invalidating the hitherto strongest argument against the role of SNAREs as fusion catalysts in this fusion step.

Concluding Remarks

Despite major progress in identifying proteins involved in fusion reactions of the endocytic pathway, we just are beginning to understand some of the underlying molecular steps and the principles that govern these reactions. The advances in imaging fusion in live cells have shown that membrane traffic is much more dynamic than previously thought, with myriads of vesicles and cisternae being constantly on the move and continuously splitting and fusing. Despite the emerging molecular complexity these reactions must be robust and adaptable. It is becoming apparent that docking and fusion steps involve assemblies of macromolecules that are not stable but rather assembled on demand and dissociate once the task is completed. Furthermore, these molecular machines appear to represent "dirty" nanostructures that do not have a fixed stoichiometry and that appear to possess a high degree of redundancy, thus being able to afford even major changes in the availability of individual components. While common principles are emerging including the stereotype action of the fusion catalysts, a lot needs to be learned about many of the other steps of the reaction cascade such as the initial signalling events and the layers of regulation that control fusion.

Acknowledgement

This work was supported by a grant from the Deutsche Forschungsgemeinschaft (SFB 523, TP B6) to R. Jahn.

References

1. Woodman PG, Warren G. Isolation of functional, coated, endocytic vesicles. J Cell Biol 1991; 112:1133-1141.
2. Mayorga LS, Diaz R, Stahl PD. Plasma membrane-derived vesicles containing receptor-ligand complexes are fusogenic with early endosomes in a cell-free system. J Biol Chem 1988; 263:17213-17216.
3. Gruenberg J, Howell KE. An internalized transmembrane protein resides in a fusion-competent endosome for less than 5 minutes. Proc Natl Acad Sci USA 1987; 84:5758-5762.
4. Gruenberg JE, Howell KE. Reconstitution of vesicle fusions occurring in endocytosis with a cell-free system. EMBO J 1986; 5:3091-3101.
5. Braell WA. Fusion between endocytic vesicles in a cell-free system. Proc Natl Acad Sci USA 1987; 84:1137-1141.
6. Itin C, Rancano C, Nakajima Y et al. A novel assay reveals a role for soluble N-ethylmaleimide-sensitive fusion attachment protein in mannose 6-phosphate receptor transport from endosomes to the trans Golgi network. J Biol Chem 1997; 272:27737-27744.
7. Cook NR, Row PE, Davidson HW. Lysosome associated membrane protein 1 (Lamp1) traffics directly from the TGN to early endosomes. Traffic 2004; 5:685-699.
8. Dunn KW, Maxfield FR. Delivery of ligands from sorting endosomes to late endosomes occurs by maturation of sorting endosomes. J Cell Biol 1992; 117:301-310.
9. Maxfield FR, McGraw TE. Endocytic recycling. Nat Rev Mol Cell Biol 2004; 5:121-132.
10. Aniento F, Emans N, Griffiths G et al. Cytoplasmic dynein-dependent vesicular transport from early to late endosomes. J Cell Biol 1993; 123:1373-1387.
11. Gruenberg J, Griffiths G, Howell KE. Characterization of the early endosome and putative endocytic carrier vesicles in vivo and with an assay of vesicle fusion in vitro. J Cell Biol 1989; 108:1301-1316.
12. Luzio JP, Rous BA, Bright NA et al. Lysosome-endosome fusion and lysosome biogenesis. J Cell Sci 2000; 113:1515-1524.
13. Mullock BM, Bright NA, Fearon CW et al. Fusion of lysosomes with late endosomes produces a hybrid organelle of intermediate density and is NSF dependent. J Cell Biol 1998; 140:591-601.
14. Pryor PR, Mullock BM, Bright NA et al. Combinatorial SNARE complexes with VAMP7 or VAMP8 define different late endocytic fusion events. EMBO Rep 2004; 5:590-595.
15. Ward DM, Leslie JD, Kaplan J. Homotypic lysosome fusion in macrophages: Analysis using an in vitro assay. J Cell Biol 1997; 139:665-673.
16. Ward DM, Pevsner J, Scullion MA et al. Syntaxin 7 and VAMP-7 are soluble N-ethylmaleimide-sensitive factor attachment protein receptors required for late endosome-lysosome and homotypic lysosome fusion in alveolar macrophages. Mol Biol Cell 2000; 11:2327-2333.
17. Abazeed ME, Blanchette JM, Fuller RS. Cell-free transport from the trans-golgi network to late endosome requires factors involved in formation and consumption of clathrin-coated vesicles. J Biol Chem 2005; 280:4442-4450.
18. Blanchette JM, Abazeed ME, Fuller RS. Cell-free reconstitution of transport from the trans-golgi network to the late endosome/prevacuolar compartment. J Biol Chem 2004; 279:48767-48773.
19. Bomsel M, Parton R, Kuznetsov SA et al. Microtubule- and motor-dependent fusion in vitro between apical and basolateral endocytic vesicles from MDCK cells. Cell 1990; 62:719-731.
20. Sönnichsen B, De Renzis S, Nielsen E et al. Distinct membrane domains on endosomes in the recycling pathway visualized by multicolor imaging of Rab4, Rab5, and Rab11. J Cell Biol 2000; 149:901-914.
21. Gruenberg J. The endocytic pathway: A mosaic of domains. Nat Rev Mol Cell Biol 2001; 2:721-730.
22. Wickner W, Haas A. Yeast homotypic vacuole fusion: A window on organelle trafficking mechanisms. Annu Rev Biochem 2000; 69:247-275.
23. Munn AL. The yeast endocytic membrane transport system. Microsc Res Tech 2000; 51:547-562.
24. Conradt B, Shaw J, Vida T et al. In vitro reactions of vacuole inheritance in Saccharomyces cerevisiae. J Cell Biol 1992; 119:1469-1479.
25. Merz AJ, Wickner WT. Resolution of organelle docking and fusion kinetics in a cell-free assay. Proc Natl Acad Sci USA 2004; 101:11548-11553.

26. Wickner W. New EMBO member's review: Yeast vacuoles and membrane fusion pathways. Embo J 2002; 21:1241-1247.
27. Mallard F, Antony C, Tenza D et al. Direct pathway from early/recycling endosomes to the Golgi apparatus revealed through the study of shiga toxin B-fragment transport. J Cell Biol 1998; 143:973-990.
28. Ghosh RN, Mallet WG, Soe TT et al. An endocytosed TGN38 chimeric protein is delivered to the TGN after trafficking through the endocytic recycling compartment in CHO cells. J Cell Biol 1998; 142:923-936.
29. Rubino M, Miaczynska M, Lippe R et al. Selective membrane recruitment of EEA1 suggests a role in directional transport of clathrin-coated vesicles to early endosomes. J Biol Chem 2000; 275:3745-3748.
30. Luzio JP, Poupon V, Lindsay MR et al. Membrane dynamics and the biogenesis of lysosomes. Mol Membr Biol 2003; 20:141-154.
31. Lippincott-Schwartz J, Roberts TH, Hirschberg K. Secretory protein trafficking and organelle dynamics in living cells. Annu Rev Cell Dev Biol 2000; 16:557-589.
32. Dumas JJ, Merithew E, Sudharshan E et al. Multivalent endosome targeting by homodimeric EEA1. Mol Cell 2001; 8:947-958.
33. Zerial M, McBride H. Rab proteins as membrane organizers. Nat Rev Mol Cell Biol 2001; 2:107-117.
34. Nichols BJ, Ungermann C, Pelham HR et al. Homotypic vacuolar fusion mediated by t- and v-SNAREs. Nature 1997; 387:199-202.
35. Ungermann C, Fischer von Mollard G, Jensen ON et al. Three v-SNAREs and two t-SNAREs, present in a pentameric cis-SNARE complex on isolated vacuoles, are essential for homotypic fusion. J Cell Biol 1999; 145:1435-1442.
36. Rieder SE, Emr SD. A novel RING finger protein complex essential for a late step in protein transport to the yeast vacuole. Mol Biol Cell 1997; 8:2307-2327.
37. Seals DF, Eitzen G, Margolis N et al. A Ypt/Rab effector complex containing the Sec1 homolog Vps33p is required for homotypic vacuole fusion. Proc Natl Acad Sci USA 2000; 97:9402-9407.
38. Sato TK, Rehling P, Peterson MR et al. Class C Vps protein complex regulates vacuolar SNARE pairing and is required for vesicle docking/fusion. Mol Cell 2000; 6:661-671.
39. Mayer A, Wickner W. Docking of yeast vacuoles is catalyzed by the Ras-like GTPase Ypt7p after symmetric priming by Sec18p (NSF). J Cell Biol 1997; 136:307-317.
40. Wang L, Seeley ES, Wickner W et al. Vacuole fusion at a ring of vertex docking sites leaves membrane fragments within the organelle. Cell 2002; 108:357-369.
41. Wang L, Merz AJ, Collins KM et al. Hierarchy of protein assembly at the vertex ring domain for yeast vacuole docking and fusion. J Cell Biol 2003; 160:365-374.
42. Stenmark H, Olkkonen VM. The Rab GTPase family. Genome Biol 2001; 2.
43. Pfeffer SR. Rab GTPases: Specifying and deciphering organelle identity and function. Trends Cell Biol 2001; 11:487-491.
44. Söllner T, Whiteheart SW, Brunner M et al. SNAP receptors implicated in vesicle targeting and fusion. Nature 1993; 362:318-324.
45. Fasshauer D, Otto H, Eliason WK et al. Structural changes are associated with soluble N-ethylmaleimide-sensitive fusion protein attachment protein receptor complex formation. J Biol Chem 1997; 272:28036-28041.
46. Antonin W, Fasshauer D, Becker S et al. Crystal structure of the endosomal SNARE complex reveals common structural principles of all SNAREs. Nat Struct Biol 2002; 9:107-111.
47. Sutton RB, Fasshauer D, Jahn R et al. Crystal structure of a SNARE complex involved in synaptic exocytosis at 2.4 A resolution. Nature 1998; 395:347-353.
48. Fasshauer D, Sutton RB, Brunger AT et al. Conserved structural features of the synaptic fusion complex: SNARE proteins reclassified as Q- and R-SNAREs. Proc Natl Acad Sci USA 1998; 95:15781-15786.
49. Fasshauer D, Antonin W, Subramaniam V et al. SNARE assembly and disassembly exhibit a pronounced hysteresis. Nat Struct Biol 2002; 9:144-151.
50. Hanson PI, Roth R, Morisaki H et al. Structure and conformational changes in NSF and its membrane receptor complexes visualized by quick-freeze/deep-etch electron microscopy. Cell 1997; 90:523-535.
51. Jahn R, Hanson PI. Membrane fusion. SNAREs line up in new environment. Nature 1998; 393:14-15.
52. Lin RC, Scheller RH. Structural organization of the synaptic exocytosis core complex. Neuron 1997; 19:1087-1094.
53. Weber T, Zemelman BV, McNew JA et al. SNAREpins: Minimal machinery for membrane fusion. Cell 1998; 92:759-772.

54. Schuette CG, Hatsuzawa K, Margittai M et al. Determinants of liposome fusion mediated by synaptic SNARE proteins. Proc Natl Acad Sci USA 2004; 101:2858-2863.
55. McNew JA, Parlati F, Fukuda R et al. Compartmental specificity of cellular membrane fusion encoded in SNARE proteins. Nature 2000; 407:153-159.
56. Melia TJ, Weber T, McNew JA et al. Regulation of membrane fusion by the membrane-proximal coil of the t- SNARE during zippering of SNAREpins. J Cell Biol 2002; 158:929-940.
57. Söllner T, Bennett MK, Whiteheart SW et al. A protein assembly-disassembly pathway in vitro that may correspond to sequential steps of synaptic vesicle docking, activation, and fusion. Cell 1993; 75:409-418.
58. Whiteheart SW, Schraw T, Matveeva EA. N-ethylmaleimide sensitive factor (NSF) structure and function. Int Rev Cytol 2001; 207:71-112.
59. Clary DO, Rothman JE. Purification of three related peripheral membrane proteins needed for vesicular transport. J Biol Chem 1990; 265:10109-10117.
60. Barnard RJ, Morgan A, Burgoyne RD. Stimulation of NSF ATPase activity by alpha-SNAP is required for SNARE complex disassembly and exocytosis. J Cell Biol 1997; 139:875-883.
61. Fleming KG, Hohl TM, Yu RC et al. A revised model for the oligomeric state of the N-ethylmaleimide-sensitive fusion protein, NSF. J Biol Chem 1998; 273:15675-15681.
62. Jahn R, Lang T, Südhof TC. Membrane fusion. Cell 2003; 112:519-533.
63. Fasshauer D, Antonin W, Margittai M et al. Mixed and noncognate SNARE complexes. Characterization of assembly and biophysical properties. J Biol Chem 1999; 274:15440-15446.
64. Yang B, Gonzalez Jr L, Prekeris R et al. SNARE interactions are not selective. Implications for membrane fusion specificity. J Biol Chem 1999; 274:5649-5653.
65. Paumet F, Brugger B, Parlati F et al. A t-SNARE of the endocytic pathway must be activated for fusion. J Cell Biol 2001; 155:961-968.
66. Atlashkin V, Kreykenbohm V, Eskelinen EL et al. Deletion of the SNARE vti1b in mice results in the loss of a single SNARE partner, syntaxin 8. Mol Cell Biol 2003; 23:5198-5207.
67. Fischer von Mollard G, Nothwehr SF, Stevens TH. The yeast v-SNARE Vti1p mediates two vesicle transport pathways through interactions with the t-SNAREs Sed5p and Pep12p. J Cell Biol 1997; 137:1511-1524.
68. McNew JA, Sogaard M, Lampen NM et al. Ykt6p, a prenylated SNARE essential for endoplasmic reticulum-Golgi transport. J Biol Chem 1997; 272:17776-17783.
69. Lewis MJ, Rayner JC, Pelham HR. A novel SNARE complex implicated in vesicle fusion with the endoplasmic reticulum. EMBO J 1997; 16:3017-3024.
70. Link E, McMahon H, Fischer von Mollard G et al. Cleavage of cellubrevin by tetanus toxin does not affect fusion of early endosomes. J Biol Chem 1993; 268:18423-18426.
71. Montecucco C, Schiavo G. Mechanism of action of tetanus and botulinum neurotoxins. Mol Microbiol 1994; 13:1-8.
72. Antonin W, Holroyd C, Fasshauer D et al. A SNARE complex mediating fusion of late endosomes defines conserved properties of SNARE structure and function. EMBO J 2000; 19:6453-6464.
73. Wang CC, Ng CP, Lu L et al. A role of VAMP8/endobrevin in regulated exocytosis of pancreatic acinar cells. Dev Cell 2004; 7:359-371.
74. Simonsen A, Gaullier JM, D'Arrigo A et al. The Rab5 effector EEA1 interacts directly with syntaxin-6. J Biol Chem 1999; 274:28857-28860.
75. McBride HM, Rybin V, Murphy C et al. Oligomeric complexes link Rab5 effectors with NSF and drive membrane fusion via interactions between EEA1 and syntaxin 13. Cell 1999; 98:377-386.
76. Dulubova I, Yamaguchi T, Gao Y et al. How Tlg2p/syntaxin 16 'snares' Vps45. EMBO J 2002; 21:3620-3631.
77. Sun W, Yan Q, Vida TA et al. Hrs regulates early endosome fusion by inhibiting formation of an endosomal SNARE complex. J Cell Biol 2003; 162:125-137.
78. Gallwitz D, Jahn R. The riddle of the Sec1/Munc-18 proteins - new twists added to their interactions with SNAREs. Trends Biochem Sci 2003; 28:113-116.
79. Toonen RF, Verhage M. Vesicle trafficking: Pleasure and pain from SM genes. Trends Cell Biol 2003; 13:177-186.
80. Misura KM, Scheller RH, Weis WI. Three-dimensional structure of the neuronal-Sec1-syntaxin 1a complex. Nature 2000; 404:355-362.
81. Bracher A, Weissenhorn W. Structural basis for the Golgi membrane recruitment of Sly1p by Sed5p. EMBO J 2002; 21:6114-6124.
82. Jahn R. Sec1/Munc18 proteins: Mediators of membrane fusion moving to center stage. Neuron 2000; 27:201-204.
83. Rizo J, Südhof TC. Snares and Munc18 in synaptic vesicle fusion. Nat Rev Neurosci 2002; 3:641-653.

84. Dulubova I, Yamaguchi T, Arac D et al. Convergence and divergence in the mechanism of SNARE binding by Sec1/Munc18-like proteins. Proc Natl Acad Sci USA 2003; 100:32-37.

85. Bryant NJ, James DE. Vps45p stabilizes the syntaxin homologue Tlg2p and positively regulates SNARE complex formation. EMBO J 2001; 20:3380-3388.

86. Nielsen E, Christoforidis S, Uttenweiler-Joseph S et al. Rabenosyn-5, a novel Rab5 effector, is complexed with hVPS45 and recruited to endosomes through a FYVE finger domain. J Cell Biol 2000; 151:601-612.

87. Ungermann C, Sato K, Wickner W. Defining the functions of trans-SNARE pairs. Nature 1998; 396:543-548.

88. Peters C, Bayer MJ, Buhler S et al. Trans-complex formation by proteolipid channels in the terminal phase of membrane fusion. Nature 2001; 409:581-588.

89. Peters C, Andrews PD, Stark MJ et al. Control of the terminal step of intracellular membrane fusion by protein phosphatase 1. Science 1999; 285:1084-1087.

90. Peters C, Mayer A. Ca2+/calmodulin signals the completion of docking and triggers a late step of vacuole fusion. Nature 1998; 396:575-580.

91. Merz AJ, Wickner WT. Trans-SNARE interactions elicit Ca2+ efflux from the yeast vacuole lumen. J Cell Biol 2004; 164:195-206.

92. Sun-Wada GH, Wada Y, Futai M. Vacuolar H+ pumping ATPases in luminal acidic organelles and extracellular compartments: Common rotational mechanism and diverse physiological roles. J Bioenerg Biomembr 2003; 35:347-358.

93. Mayer A. Membrane fusion in eukaryotic cells. Annu Rev Cell Dev Biol 2002; 18:289-314.

94. Starai VJ, Thorngren N, Fratti RA et al. Ion regulation of homotypic vacuole fusion in Saccharomyces cerevisiae. J Biol Chem 2005; 280:16754-16762.

95. Boeddinghaus C, Merz AJ, Laage R et al. A cycle of Vam7p release from and PtdIns 3-P-dependent rebinding to the yeast vacuole is required for homotypic vacuole fusion. J Cell Biol 2002; 157:79-89.

96. Cheever ML, Sato TK, de Beer T et al. Phox domain interaction with PtdIns(3)P targets the Vam7 t-SNARE to vacuole membranes. Nat Cell Biol 2001; 3:613-618.

97. Antonin W, Holroyd C, Tikkanen R et al. The R-SNARE endobrevin/VAMP-8 mediates homotypic fusion of early endosomes and late endosomes. Mol Biol Cell 2000; 11:3289-3298.

98. Kreykenbohm V, Wenzel D, Antonin W et al. The SNAREs vti1a and vti1b have distinct localization and SNARE complex partners. Eur J Cell Biol 2002; 81:273-280.

99. Mallard F, Tang BL, Galli T et al. Early/recycling endosomes-to-TGN transport involves two SNARE complexes and a Rab6 isoform. J Cell Biol 2002; 156:653-664.

100. Tai G, Lu L, Wang TL et al. Participation of the syntaxin 5/Ykt6/GS28/GS15 SNARE complex in transport from the early/recycling endosome to the trans-golgi network. Mol Biol Cell 2004; 15:4011-4022.

101. Ungermann C, Nichols BJ, Pelham HR et al. A vacuolar v-t-SNARE complex, the predominant form in vivo and on isolated vacuoles, is disassembled and activated for docking and fusion. J Cell Biol 1998; 140:61-69.

102. Dilcher M, Kohler B, Fischer von Mollard G. Genetic interactions with the yeast Q-SNARE VTI1 reveal novel functions for the R-SNARE YKT6. J Biol Chem 2001; 276:34537-34544.

CHAPTER 6

Clathrin Adaptor Proteins in Cargo Endocytosis

Linton M. Traub*

Abstract

Eukaryotic cells continuously remodel the protein and lipid composition of the plasma membrane in response to the extracellular milieu. Membrane retrieval typically involves inward budding of small bilayer-encapsulated vesicles that shuttle protein and lipid from the surface to internal endosomal elements. Clathrin-mediated endocytosis is a dominant pathway for internalization in many cell types, and a range of dedicated signals are used to ensure selective sorting in this pathway. Evidence suggests that the clathrin coat utilizes a diverse collection of clathrin-associated sorting proteins (CLASPs) to ensure the efficient and noncompetitive concentration of a wide variety of sorting signals within transport vesicles forming at the cell surface.

Introduction

The limiting membrane of eukaryotic cells represents the primary interface with the exterior environment. It performs a vital barrier function that, by tightly controlling the entry of molecules into the cell interior, contributes to long-term cell survival. The plasma membrane is also a platform for first detecting and then responding to extracellular signals, ions and nutrients, as well as potential pathogens. This is because a large array of receptors, ion channels, pumps and transporter proteins are positioned at the cell surface of a typical eukaryotic cell. Some receptors (mostly handling nutrient uptake or protein delivery) flux constantly through the plasma membrane while others, primarily involved in signal transduction, only exit the cell surface en masse following ligand stimulation. The process of regulated removal of certain receptors from the surface is perhaps best understood from the point of development,[1] where failure to quantitatively clear a particular receptor(s) from the plasma membrane can ultimately result in inappropriate cell fate determination due to erroneous cellular responses to local morphogens or growth factors.[2,3]

Sorting Signals for Selective Internalization

In principle, internalization of transmembrane proteins (along with bound ligands) could be either an active or a passive process. There are certainly examples of proteins being retained at the plasma membrane through stabilizing interactions with lipids,[4] lipid rafts,[5] or other proteinaceous components, like scaffolding proteins with PDZ domains for example.[6] Yet, it is generally accepted that endocytic uptake is governed by a positive signal;[7] in other words, there

*Correspondence Author: Linton M. Traub—Department of Cell Biology and Physiology, University of Pittsburgh School of Medicine, 3500 Terrace Street, S325BST Pittsburgh, Pennsylvania 15261, U.S.A. Email: traub@pitt.edu

Endosomes, edited by Ivan Dikic. ©2006 Landes Bioscience and Springer Science+Business Media.

is a process of preferential sorting of select transmembrane proteins for enrichment within forming carrier vesicles. This is because many proteins/receptors stagnate at the cell surface, redistributing into a diffuse pattern over the plasma membrane, when sorting signals within the cytosolic domain are inactivated either by inherited defects or directed mutation(s), or if the cytosolic region is simply truncated.[8,9] Indeed, numerous, structurally discrete sorting signals that specify rapid internalization are now known[7] (Table 1).

Clathrin-Mediated Endocytosis

There are several ways in which preferential entry of transmembrane proteins and bound ligands into the cytoplasm can occur. These include the budding of small (~60-100 nm) membrane bound vesicles in caveolin-dependent,[10] clathrin-dependent,[11,12] and uncoated caveolin-independent endocytosis,[13,14] as well as through larger forms of membrane internalization vehicle, as in macropinocytosis and phagocytosis.[12] Of these, probably the best characterized at the mechanistic level is clathrin-mediated endocytosis. Clathrin-coated buds and clathrin-coated vesicles were originally discovered in thin-section electron micrographs on the basis of a highly characteristic bristle-like matrix deposited on the cytosolic aspect of these structures[15] (see Fig. 1A). The unusual membrane invaginations also displayed striking concentration of electron dense material within the lumen, located opposite to the bristle coat.[15] These landmark morphological findings accurately mirror the two principal functional activities of clathrin-coated vesicles (and coat-dependent sorting in general); assembly of a polymeric, cytosol-oriented coat while simultaneously gathering select cargo molecules into the invaginating coated vesicle.

Table 1. Clathrin-dependent endocytic sorting signals

Signal Type	Examples Sequence	Examples Protein	Recognition Protein/Domain	References
YXXØ[a]	YTRF	transferrin receptor	AP-2, μ2 subunit	7, 19
	YSKV	CI[b]-MPR		
	YRGV	CD[c]-MPR		
	YQRL	TGN38/46		
	YQTI	LAMP-1		
	YATL	LRP1		
[DE]XXXL[LI]	ERAPLI	LIMP-II	AP-1, γ/σ1	7, 47
	DKQTLL	CD3-γ	hemicomplex	
	EKQPLL	tyrosinase	AP-2, α/σ2	
	DQRDLI	Ii	hemicomplex?	
[FY]XNPXY	FDNPVY	LDL receptor	ARH, Dab2, Numb	7, 19
	FTNPVY	LRP1	PTB domain	
	FENPMY	megalin	β-arrestin?	
	YTNPAF	Sanpodo		
Phosphorylation	–	GPCRs	β-arrestin 1/2	19, 64
Ubiquitin	–	EGFR	epsin, eps15	102, 103
		Notch	UIM[e] domain?	
		Delta		
		ENaC[d]		

[a] Consensus sequences indicated in single letter amino acid notation using PROSITE syntax. Ø indicates a bulky hydrophobic amino acid; Leu, Met, Ile, Phe, Val. [b] CI-MPR is cation-independent mannose 6-phosphate receptor. [c] CD-MPR is cation-dependent mannose 6-phosphate receptor. [d]ENaC is epithelial sodium channel. [e]UIM is ubiquitin interaction motif.

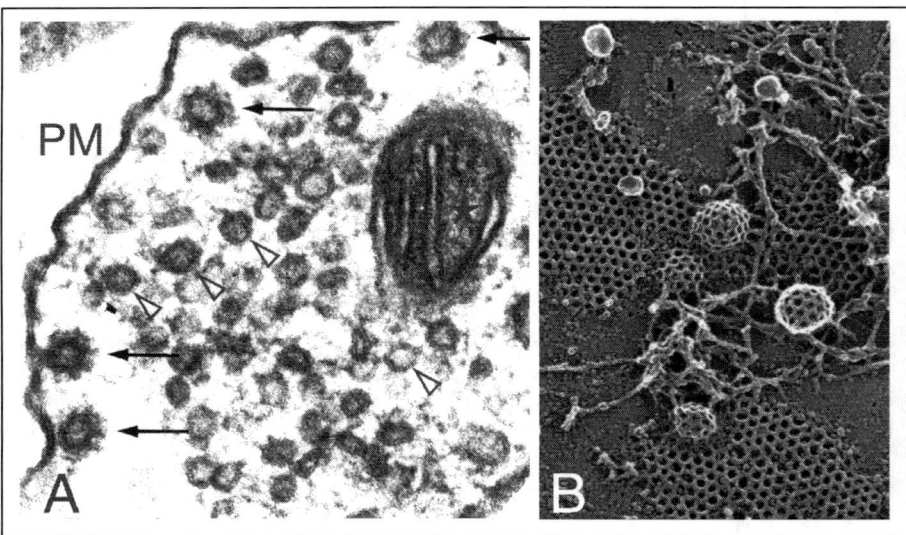

Figure 1. Clathrin-coat morphology. A) Thin-section electron micrograph of a purified rat brain nerve ending (synaptosome) revealing the characteristic cytosol-oriented 'bristle' matrix (arrows) enveloping several clathrin-coated vesicles budding from the presynaptic plasma membrane (PM). At the nerve ending, coated vesicles retrieve synaptic vesicle membrane components to replenish the pool of vesicles (arrowheads) required for sustained neurotransmission. A mito-chondrion (asterisk) is positioned in close proximity to the synaptic region. B) Freeze-etch electron micrograph of the cytoplasmic face of the ventral plasma membrane of a cultured cell revealing the typical polyhedral clathrin lattice. The progressive invagination of coated mem-brane is clearly seen in different intermediates of the budding process. This image was graciously provided by John Heuser.

The 200 Å bristle-like structures radiating off clathrin-coated membranes correspond to the regular coat assemblage. The major constituent is, of course, clathrin, a tri-legged protein complex composed of three 192-kDa heavy chains and three ~25-kDa regulatory light chains, one bound to each heavy chain[16] (Fig. 2). Clathrin trimers can self-assemble onto spherical polyhedral assemblies in vitro, which are highly reminiscent of coated vesicles in vivo. It is the geometric nature of the triskelion that dictates the formation of the hexagon/pentagon-containing lattice, the most recognizable feature of the clathrin coat (Fig. 1B). The orientation and rigidity of the fibrous heavy chains imparts an inherent sidedness to the triskelion; when assembled upon biological membranes, the amino-terminal region of the heavy chain, which folds into a 7-bladed β-propeller structure,[17] is oriented closest to the bilayer. The carboxyl termini, which bundle together at the central vertex in a helical tripod arrangement, are positioned furthest from the membrane surface.[16] However, clathrin does not have the capacity to associate with phospholipid bilayers directly. Instead, adaptor proteins couple the clathrin lattice with the underlying membrane by synchronously binding to the clathrin terminal domain, to phospholipids, and to cargo sorting signals.

The AP-2 Adaptor Complex

A second major protein constituent of the endocytic clathrin coat is the AP-2 adaptor,[18] composed of four distinct subunits; two large ~100-kDa subunits (α and $\beta2$), a 50-kDa medium subunit ($\mu2$), and a 17-kDa small subunit ($\sigma2$).[19] The functional complex has a characteristic architecture of two small appendages projecting off a larger globular core through

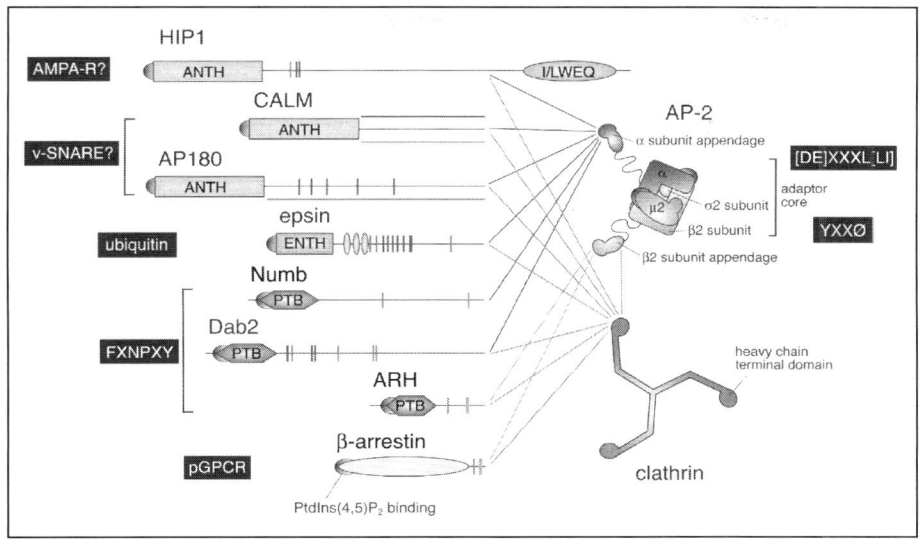

Figure 2. The clathrin–AP-2–CLASP interaction network. Schematic depiction of the protein–protein interactions between clathrin, AP-2 and selected CLASPs is shown. Images are modeled on the known molecular structures of the coat components, but are not drawn to scale. The clathrin terminal domain and the AP-2 appendages serve as platforms to coordinate the protein–protein interaction networks established during coat assembly. The location of interaction motifs used to engage these binding platforms are indicated by color-coded vertical lines. In AP180 and CALM, where the interaction motifs have not been defined precisely, interaction is generally signified by a horizontal line. The location of the three tandem UIMs in epsin 1 (ovals) is shown. pGPCR indicates phosphorylated (ligand-activated) G protein-coupled receptor, while I/LWEQ is a modular F-actin binding fold that is also conserved in the S. cerevisiae HIP1 orthologue Sla2p.

protease-sensitive stalks of apparently variable length.[20] Structural studies show the core is a heterotetramer comprised of the amino-terminal trunk regions of the large chains complexed with the smaller subunits[21] (see Fig. 2). The two appendages correspond to the independently folded carboxy-terminal segments of the large α and $\beta2$ subunits,[22-24] each attached to the respective trunk through an intrinsically unstructured intervening polypeptide hinge (for a comprehensive review of adaptor structure, see ref. 19). A five residue interaction motif (LLNLD; termed the clathrin box) is positioned within the pliable disordered hinge of the $\beta2$ subunit[25] and this binds directly to the clathrin terminal domain β-propeller.[26]

The AP-2 core complex seems to oversee coat operations close to the membrane. A patch of basic amino acids near the amino terminus of the α subunit binds to phosphoinositides, specifically phosphatidylinositol 4,5-bisphosphate (PtdIns(4,5)P$_2$).[21,27,28] A second PtdIns(4,5)P$_2$-contact site is located on the $\mu2$ subunit but, in the resting (cytosolic) state, the two chains are not positioned appropriately to allow simultaneous lipid binding. Importantly, the $\mu2$ subunit also binds to a common class of sorting signal, the tyrosine-based YXXØ motif (where X is any amino acid and Ø represents a residue with a bulky hydrophobic side chain), found for example in the transferrin receptor, the mannose 6-phosphate receptors and the low density lipoprotein (LDL) receptor-related protein 1 (LRP1)[7] (Table 1). Classical models for clathrin-based sorting do not accommodate phospholipid support, and posit simply that coat assembly proceeds via the concerted linking of cargo with clathrin through the bifunctional AP-2 adaptor to fabricate a polymeric coat that progressively invaginates before budding into the cytoplasm. The final scission step is a complex process, handled by the large GTPase dynamin and several accessory proteins such as amphiphysin, endophilin and sorting nexin 9.[29-31] Upon release of the transport vesicle, rapid, ATP-dependent uncoating allows both AP-2 and clathrin to reenter a soluble pool to initiate new rounds of coat assembly, while the vesicle is primed for fusion with acceptor endosomal elements.

But the structure of the AP-2 core reveals that in the basal state the surface of the μ2 subunit that recognizes the YXXØ sequence is clearly inaccessible.[21] This region of μ2 packs up against the adjacent β2 trunk and phosphorylation of μ2 at Thr156 is necessary to unleash the subunit for productive interactions with YXXØ sequences.[32,33] The repositioning of phosphorylated μ2 allows simultaneous binding of the YXXØ signal and PtdIns(4,5)P$_2$[21,34] strengthening the association of AP-2 with the plasma membrane. In addition, the catalytic activity of the kinase that phosphorylates Thr156, termed AAK1, is stimulated by assembled clathrin.[35,36] This suggests that AP-2 is focally activated to engage cargo as the coat assembles, possibly making cargo recognition a relatively late event. Indeed, μ2 subunit mutations that preclude YXXØ engagement do not prevent the proper deposition of AP-2 on the plasma membrane.[32,37] Nonetheless, cargo capture does appear to be a decisive step in the assembly process as live-cell imaging of clathrin dynamics in cultured cells shows that failure to concentrate cargo (transferrin) within an assembling lattice rapidly triggers catastrophic dissolution of the assemblage.[38]

Mutation or targeted gene disruption of individual AP-2 subunits is lethal in mice, *Drosophila melanogaster*[39] and *Caenorhabditis elegans*[40] (although not in *Saccharomyces cerevisiae*).[41,42] In cultured mammalian cells, post-transcriptional silencing of AP-2 α or μ2 subunit mRNA using small interfering RNA (siRNA) halts the internalization of the transferrin receptor[43-45] and reduces the number of morphologically discernable (bristle-like) clathrin coats at the surface more than tenfold.[44] These results clearly highlight the pivotal contribution AP-2 makes to clathrin-mediated endocytosis.

Diversity in Cargo Recognition Events

AP-2 is structurally and functionally homologous to the AP-1 and AP-3 heterotetramers that operate at the trans-Golgi network and on endosomes.[46] A nontyrosine sorting signal, the [DE]XXXL[LI]-type dileucine sequence (Table 1) binds to a hemicomplex of the γ1/σ1 or δ/σ3 subunits of AP-1 or AP-3, respectively, at a presently unknown location.[47] The [DE]XXXL[LI] sequence also acts as an efficient internalization motif[48] so it probable that AP-2 binds physically to this sorting signal as well. The capability of AP-2 to contact two discrete signals though separate subunits implies that individual coated vesicles can simultaneously carry multiple classes of cargo. In reality, endocytic clathrin coats do not only garner YXXØ- and [DE]XXXL[LI]-harboring proteins. Other receptors with tyrosine-based FXNPXY-type internalization sequences, like the LDL receptor, likewise cluster at assembled clathrin structures at the cell surface.[49-51] And still others, like the epidermal growth factor (EGF) receptor, also congregate in clathrin-coated regions following activation.[49,52] Although the EGF receptor contains a YXXØ signal (YRAL) within the cytosolic domain and binds directly to AP-2,[53] this segment is dispensable for internalization.[54] Accordingly, point mutations within the μ2 subunit that prevent YXXØ-sequence binding have little inhibitory effect on EGF receptor uptake.[37] In fact, ectopic receptor overexpression studies show unambiguously that saturating levels of FXNPXY or [DE]XXXL[LI]-containing proteins at the surface have negligible effect on the kinetics of transferrin or EGF receptor internalization and *vice versa*.[48,55,56] And each receptor type saturates the endocytic machinery at different surface densities.[56] One explanation for these general findings could be that the internalization of each cargo type is overseen by distinct sets of clathrin-coated vesicle with nonoverlapping cargo selectivity. Ultrastructural analysis of surface clathrin-coated regions argue strongly against this notion. In electron micrographs, receptors for transferrin, LDL, EGF and insulin and are clearly colocalized within individual coated profiles;[49,57] each clathrin-coated vesicle contains numerous types of cargo utilizing different sorting signals which, moments later, are found in common early endosomes.[58] Good coincidence of the transferrin and LDL receptor in individual punctate

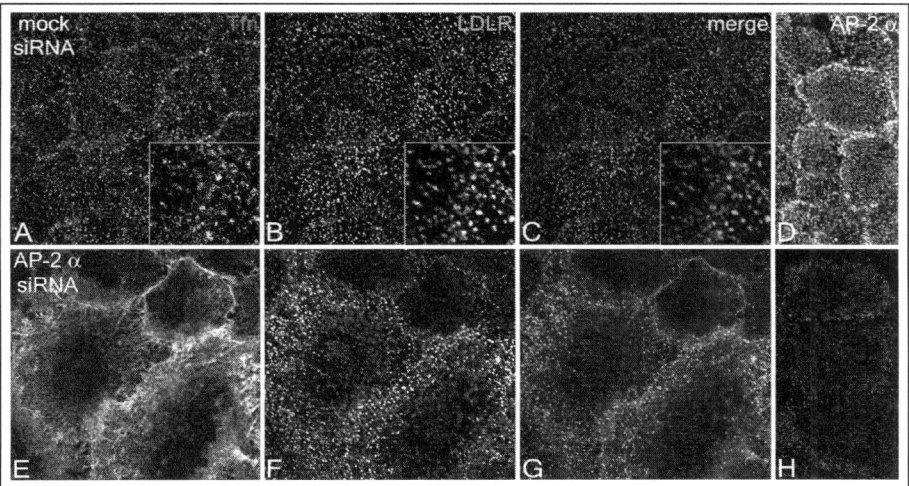

Figure 3. AP-2 does not regulate the sorting of all clathrin-dependent cargo. Single confocal optical sections of confluent HeLa cell monolayers treated either with mock (A-D) or AP-2 α subunit siRNA (E-H) and then double-labeled with fluorescent transferrin (A and E, red) and anti-LDL receptor antibodies (B and F, green), or stained with anti-AP-2 α subunit antibodies (D and H). In untreated cells, the transferrin and LDL receptors colocalize but after AP-2 silencing, only the transferrin receptor undergoes massive redistribution at the cell surface. Insets in A, B, and C show a magnified region to better judge the degree of colocalization between transferrin and the LDL receptor.

(clathrin-containing) sites on the plasma membrane is also seen by light microscopy (Fig. 3A-D). But, while RNAi silencing of the AP-2 α subunit (Fig. 3H) causes the transferrin receptor to diffuse over the cell surface (Fig. 3E), the LDL receptor, strikingly, remains concentrated in punctate zones (Fig. 3F) and internalizes rather efficiently, even in the absence of AP-2.[44] Entry of the LDL receptor is still clathrin dependent under these conditions because siRNA-mediated knock down of clathrin prevents the internalization of both transferrin and LDL receptors.[44] This capability of receptors lacking a YXXØ -or [DE]XXXL[LI]-type internalization signal to enter the cell in the absence of functional AP-2 suggests that alternate adaptor proteins must regulate the internalization of some classes of cargo and these might correspond to the differentially saturable connectors revealed in the overexpression studies.[48,56]

Clathrin-Associated Sorting Proteins (CLASPs), the Alternate Adaptors

Early clues to the nature of putative alternate adaptors came from studies on the seven-transmembrane-spanning G protein-coupled receptors (GPCRs). GPCRs are the most extensive superfamily of signaling receptors in eukaryotes and many, but not all, of these proteins are quickly downregulated by clathrin-mediated endocytosis following agonist-induced activation.[59-61] Ligand binding prompts phosphorylation of the cytosolic region of most GPCRs. This reversible modification acts as a trigger for one of two ubiquitous arrestin family members (β-arrestin 1 or β-arrestin 2) to translocate onto the activated GPCR.[59-61] The β-arrestins desensitize the receptor by uncoupling signaling events, but also direct the phosphorylated GPCR to clathrin-coated structures for prompt internalization. The capacity of β-arrestins to promote GPCR endocytosis depends upon four fundamental properties of these adaptor proteins: the ability to bind physically to the GPCR, to PtdIns(4,5)P$_2$, to AP-2 and to clathrin.[62-64] Interaction motifs arrayed tandemly within the carboxy-terminal segment of the β-arrestins specify direct interactions with the clathrin heavy chain β-propeller[26,65,66] and with the appendage of the AP-2 β2 subunit.[63] All four interactions are required for efficient GPCR endocytosis, and β-arrestins work by melding the GPCR with preexisting clathrin coats by synchronously engaging both.[67] These are essentially the properties of the archetypical adaptor, AP-2, but it is the β-arrestins, and not AP-2 directly, that are charged with overseeing the internalization of the largest family of receptors known. There is also evidence that β-arrestin 2 handles the uptake of nonclassical seven-transmembrane-spanning smoothened[68] and frizzled[69] receptors that operate in the Wnt and Hedgehog signaling pathways.[1] Also, β-arrestin 2 apparently participates in the internalization of transforming growth factor β (TGFβ) by binding to the type III TGFβ receptor (also termed betaglycan) in a phosphorylation-sensitive manner.[70] And finally, β-arrestin nullizygous mice fed a high fat diet display defective lipoprotein metabolism and it appears that β-arrestin 2 may also modulate constitutive LDL receptor endocytosis by binding to the cytosolic domain of the receptor in a phosphorylation-dependent fashion.[71]

The cargo selective properties of the β-arrestins make them the founding members of the CLASP family. But certain other so-called endocytic 'accessory' proteins[72] also display the important binding properties typical of the β-arrestins (and AP-2).[73] These presumptive CLASPs bind to PtdIns(4,5)P$_2$, albeit using different modular lipid-binding folds, bind to the clathrin β-propeller domain, and all can engage the appendages of the large AP-2 adaptor subunits[19,46,73] (Fig. 2). CLASPs appear to operate by contributing to the efficiency of clathrin lattice assembly while simultaneously expanding the sorting repertoire of the forming coat.[46,74] Currently, the experimental evidence for a set of dedicated, phosphotyrosine-binding (PTB) domain-containing CLASPs governing LDL receptor internalization is most compelling.

PTB Domain CLASPs and FXNPXY Signal Endocytosis

The PTB acronym is actually a misnomer. Although the domain was originally identified as a non-SH2 phosphotyrosine binding module,[75] it is clear that tyrosine phosphorylation is not a prerequisite for productive binding of the canonical PTB binding-partner sequence [FY]XNPXY.[76] A form of this sequence positioned within the cytosolic domain of the LDL receptor (FDNPVY), in its nonphosphorylated form, was actually the first internalization signal ever discovered.[77] Two PTB domain CLASPs, designated Disabled-2 (Dab2) and the autosomal recessive hypercholesterolemia (ARH) protein, appear to decode the FXNPXY internalization signal in the LDL and related receptors. Inherited mutations or targeted gene disruption of these proteins cause obvious pathological phenotypes.[78-80] For ARH, individuals with inactivating mutations in both *ARH* alleles have early onset hypercholesterolemia that is very difficult to distinguish from that typical of patients entirely lacking LDL receptor function.[78,81-86] Thus, these unfortunate individuals present with a disease that closely mirrors a complete failure to bind and clear circulating LDL particles. In full agreement, ARH$^{-/-}$ mice show dramatic accumulation of LDL receptors at the sinusoidal plasma membrane of hepatocytes, the cell type responsible for clearing the bulk of circulating LDL from the plasma.[87]

In a similar fashion, Dab2 nullizygous mice display a proteinuria because of ineffective sorting of an LDL superfamily receptor termed megalin at the apical surface of renal proximal tubules.[79,88] Megalin is a scavenger receptor that retrieves considerable amounts of albumin, vitamin- and lipid-binding proteins from the glomerular filtrate to prevent protein excretion in the urine;[79] the Dab2$^{-/-}$ phenotype is actually a milder version of that seen upon megalin gene disruption.[89] ARH and Dab2 use a single folded PTB domain to bind the FXNPXY sequence and PtdIns(4,5)P$_2$ noncompetitively.[51,90-92] This promotes cargo recognition, while binding to the clathrin coat is controlled by the carboxy-terminal segment in both proteins.[51,90,93,94] Just like the β-arrestins, ARH and Dab2 have tandemly arrayed interaction sequences that promote associations with the AP-2 appendages and the clathrin β-propeller (Fig. 2). That these proteins can contribute to clathrin coat assembly is evidenced by a decrease in detectable clathrin-coated vesicles at the apical pole of Dab2-null proximal tubules.[79]

Secondary structure predictions suggest that outside of the modular amino-terminal PTB domain, these proteins are largely unstructured. The reason for this may be that intrinsically disordered tracts of polypeptide are highly mobile and pliable, which allows a large 'capture radius' for binding partners. This property could draw receptor–CLASP complexes into a pre-existing clathrin structure rather efficiently. A gross absence of secondary structural elements is also consistent with what is known about the mode of engagement of various interaction motifs with the clathrin β-propeller and the AP-2 appendages.[26,95-97] In fact, extended regions of unstructured random-coil polypeptide turns out to be an unexpected characteristic of the CLASP group of endocytic components.[73,98]

Another very informative example of the endocytic action of a PTB-domain CLASP, termed Numb, is in the biogenesis of retro-orbital neurosensory bristles in *Drosophila*.[99] The mature sensory bristle is composed of four cell types; hair, socket, neuron and sheath cells. The four all arise from a single sensory organ precursor (SOP) cell by two rounds of asymmetric cell division.[3,100] During SOP cell mitosis, Numb distributes asymmetrically in the mother cell so as to selectively partition into only one of the two daughter cells.[100] Mutant *numb* alleles lead to cell fate transformations, resulting in only four socket cells being formed.[3,100] Mutations that disrupt the activity of the appendage of the AP-2 α subunit, to which Numb binds directly, generate a phenotype equivalent to *numb* mutants.[3] The simplest model for Numb activity is that asymmetric Numb–AP-2 association during cell division results in one of the two siblings having a considerably higher Numb/AP-2 complement, which then drives the removal of the transmembrane Notch receptor from the surface in this cells. This, then, changes the fate of the cell compared to the other daughter.[3,99] Numb binds to Notch, yet there are no obvious differences in Notch expression level between the two cells that ultimately become quite different cell types. Instead, a Notch-binding accessory protein termed Sanpodo, seems to be the surface component downregulated in Numb-enriched daughter cells.[101] Sanpodo is a tetraspannin membrane protein with a variant [YF]XNPXY sequence (YTNPAF) in the cytosolic portion and Numb-mediated clearance of Sanpodo from the surface is suggested to compromise Notch signaling, leading to a different cell fate.[101]

ENTH/ANTH-Domain Containing CLASPs

Posttranslational conjugation of ubiquitin is another important method of generating sorting signals, as covered in several other chapters of this book (see chapters by Puertollano, Pattni and Stenmark, and Gur et al). A discussion of the nature and generation of this internalization signal is beyond the scope of this chapter, but the epsin family of CLASPs (together with eps15, an epsin binding partner) are good candidates for recognizing ubiquitinated cargo at the cell surface.[102,103] Again, the conserved modular architecture and binding properties of epsin are fully compatible with a role as a cargo-selective adaptor. The epsin N-terminal homology (ENTH) domain binds directly to PtdIns(4,5)P$_2$,[104] while two or three ubiquitin-interaction motifs (UIMs) enable the protein to bind physically to ubiquitin[105] (Fig. 2). The UIMs are located between the ENTH domain and the unstructured carboxy-terminal region, housing standard interaction motifs that enable epsin to bind both clathrin and the AP-2 appendages[106-108] (Fig. 2).

In *Drosophila*, while a PTB-domain CLASP (Numb) attenuates Notch receptor signal transduction in certain SOP progeny, an ENTH-domain CLASP (epsin) actively promotes Notch signaling.[109-112] Notch is activated in a signal-receiving cell by engaging the transmembrane ligand Delta, presented at the surface of an adjacent signal-sending cell.[113] The Notch transcriptional response is preceded by two ordered, ligand-induced cleavage events that sever Notch on both sides of the plasma membrane allowing the intracellular domain to translocate to the nucleus.[113] It appears that the second proteolytic event that liberates the Notch intracellular domain requires prior endocytosis of the extracellular region of Notch, bound to Delta, into the signal-sending cell. There are strong genetic interactions between Delta, two E3 ubiquitin ligases (Neuralized and Mind bomb), and the fly epsin, termed Liquid facets.[109-112] Current evidence strongly suggests that the E3-mediated ubiquitination of Delta that allows Liquid facets to internalize the Delta/Notch complex is obligatory for Notch signaling. The pivotal role of Liquid facets in promoting Notch signaling is underscored by the participation of the deubiquitinating enzyme Fat facets. Fat facets appears to operate by buffering the intracellular epsin concentration by salvaging proteasome destined, polyubiquitinated Liquid facets by deubiquitinating the protein.[114] In fact, expression of a single extra copy of Liquid facets abolishes the requirement for Fat facets in the developing compound eye; there is no eye phenotype in Fat facets-null flies expressing an extra Liquid facets gene.[2] The *Drosophila* data suggest that during development, the intracellular abundance of epsin is critical, presumably to effect the timely endocytosis of certain membrane proteins/receptors thereby facilitating appropriate cell fate determination.[2,110,113] This is in full accord with the concept of epsin functioning as a CLASP. Likewise, in mammals, another ENTH domain protein, huntingtin interacting protein 1 (HIP1) displays a variation on the general CLASP architecture (Fig. 2), and may manage the internalization of AMPA receptors in the central nervous system.[115]

Finally, the structure of the AP180 N-terminal homology (ANTH) domain is highly related to the ENTH module and binds to PtdIns(4,5)P$_2$, although by a different molecular mechanism.[19] AP180 and the nonneuronal orthologue CALM both bind clathrin and AP-2. As for function, UNC-11, the *C. elegans* AP180, is abundantly expressed in neurons and analysis of the effect of unc-11 mutants suggests that this protein regulates the retrieval of the v-SNARE synaptobrevin from the presynaptic plasma membrane.[116]

Conclusions and Perspective

The identification of cargo selective CLASPs allows several previously unexplained findings now to be resolved. General collaboration between AP-2 and CLASPs can adequately explain how PtdIns(4,5)P$_2$ within the cytosolic leaflet of the plasma membrane can act as a compartmental cue to direct the assembly of endocytic clathrin coats. The overall architectural similarity of the CLASP family is striking (Fig. 2), and the capability of numerous CLASPs to bind physically to both AP-2 and clathrin could account for the very rapid assembly of coated vesicles and the swift internalization of a diverse array of transmembrane proteins from the cell surface. Perhaps most importantly, the activity of these proteins can explain how internalization of select transmembrane proteins can continue in cells essentially lacking functional AP-2.[43-45]

Intriguingly, vesicle coats that operate at other sorting stations within the cell (the COPI and COPII coats, for example) do not appear to have diversified cargo capture operations to involve a host of CLASP-like components.[73,117] Thus a legitimate question is why there are so many endocytic CLASPs (and we may not yet have the complete list of cargo specific components operating at the cell surface). As discussed above, in *Drosophila*, the Notch pathway utilizes both PTB- (Numb) and ENTH- (Liquid facets) domain CLASPs to fine-tune signaling activity. But Notch/Sanpodo and the ligand Delta are endocytosed in different cells, the signal-receiving and signal-sending cell, respectively. So one obvious possibility is that the complexity of the endocytic CLASP network (Fig. 2) is substantially reduced in individual cell types. In other words, perhaps not all coated structures contain a full complement of CLASPs. This seems unlikely, however, because in SOP cell progeny for example, the daughter cell utilizing Numb to suppress intracellular Notch signal transduction is simultaneously using Liquid facets to promote effective Notch signaling in the adjacent sister cell. In fact, the vast majority of the core CLASP members are found both in brain and in cultured lines, like HeLa cells and fibroblasts. It therefore seems more probable that numerous different CLASPs populate a single clathrin structure assembling at the surface. The synchronous operation of a diverse group of CLASPs has the clear advantage of equipping individual clathrin coats with the capability of capturing an extended range of cargo molecules at the cell surface.

But even with the incorporation of CLASP activity into models for clathrin-based sorting, our general understanding of the molecular events that underpin coat assembly is still rather rudimentary. A major challenge is now to begin to understand the temporal chronology and regulation of the numerous protein–protein interactions necessary for the successful fabrication of a clathrin-coated vesicle. While CLASPs clearly bind to the AP-2 appendages, there exists a diverse set of appendage binding sequences and two spatially separate contact sites upon each appendage.[19,73,118] Unfortunately, the static representation of dense network connectivity (Fig. 2) cannot adequately explain the assembly and cargo selection process chronologically; we do not understand the flow and processing of information. For instance, the biologic advantage of one CLASP displaying a particular set of appendage-binding sequences and not another is not at all clear. Presumably these sequence dictate the temporal behavior and possibly the rank of a particular CLASP, but there is a paucity of functional information on the individual contributions of these interactions to the coupled process of sorting and budding. It seems certain that the application of a wide array of experimental approaches, including live-cell imaging,[38,119] RNAi and gene replacement, time-resolved proteomics[120] and, perhaps most importantly, computational analysis and network theory modeling[121] will be necessary to unlock all the secrets of this highly efficient and sophisticated vesicular shuttle.

References

1. Le Roy C, Wrana JL. Clathrin- and nonclathrin-mediated endocytic regulation of cell signalling. Nat Rev Mol Cell Biol 2005; 6:112-126.
2. Cadavid AL, Ginzel A, Fischer JA. The function of the Drosophila fat facets deubiquitinating enzyme in limiting photoreceptor cell number is intimately associated with endocytosis. Development 2000; 127:1727-1736.
3. Berdnik D, Torok T, Gonzalez-Gaitan M et al. The endocytic protein α-Adaptin is required for Numb-mediated asymmetric cell division in Drosophila. Dev Cell 2002; 3:221-231.
4. Resh MD. Membrane targeting of lipid modified signal transduction proteins. Subcell Biochem 2004; 37:217-232.
5. Roper K, Corbeil D, Huttner WB. Retention of prominin in microvilli reveals distinct cholesterol-based lipid micro-domains in the apical plasma membrane. Nat Cell Biol 2000; 2:582-592.
6. Brone B, Eggermont J. PDZ proteins retain and regulate membrane transporters in polarized epithelial cell membranes. Am J Physiol Cell Physiol 2005; 288:C20-C29.
7. Bonifacino JS, Traub LM. Signals for sorting of transmembrane proteins to endosomes and lysosomes. Annu Rev Biochem 2003; 72:395-447.
8. Sanan DA, Van der Westhuyzen DR, Gevers W et al. The surface distribution of low density lipoprotein receptors on cultured fibroblasts and endothelial cells. Ultrastructural evidence for dispersed receptors. Histochemistry 1987; 86:517-523.
9. Lobel P, Fujimoto K, Ye RD et al. Mutations in the cytoplasmic domain of the 275 kd mannose 6-phosphate receptor differentially alter lysosomal enzyme sorting and endocytosis. Cell 1989; 57:787-796.
10. Nabi IR, Le PU. Caveolae/raft-dependent endocytosis. J Cell Biol 2003; 161:673-677.
11. Brodsky FM, Chen CY, Knuehl C et al. Biological basket weaving: Formation and function of clathrin-coated vesicles. Annu Rev Cell Dev Biol 2001; 17:517-568.
12. Conner SD, Schmid SL. Regulated portals of entry into the cell. Nature 2003; 422:37-44.
13. Damm EM, Pelkmans L, Kartenbeck J et al. Clathrin- and caveolin-1-independent endocytosis: Entry of simian virus 40 into cells devoid of caveolae. J Cell Biol 2005; 168:477-488.
14. Kirkham M, Fujita A, Chadda R et al. Ultrastructural identification of uncoated caveolin-independent early endocytic vehicles. J Cell Biol 2005; 168:465-476.
15. Roth TF, Porter KR. Yolk protein uptake in the oocyte of the mosquito Aedes aegypti. L J Cell Biol 1964; 20:313-332.
16. Fotin A, Cheng Y, Sliz P et al. Molecular model for a complete clathrin lattice from electron cryomicroscopy. Nature 2004; 432:573-579.
17. ter Haar E, Musacchio A, Harrison SC et al. Atomic structure of clathrin: A β propeller terminal domain joins an α zigzag linker. Cell 1998; 95:563-573.
18. Blondeau F, Ritter B, Allaire PD et al. Tandem MS analysis of brain clathrin-coated vesicles reveals their critical involvement in synaptic vesicle recycling. Proc Natl Acad Sci USA 2004; 101:3833-3838.
19. Owen DJ, Collins BM, Evans PR. Adaptors for clathrin coats: Structure and function. Annu Rev Cell Dev Biol 2004; 20:153-191.

20. Heuser JE, Keen J. Deep-etch visualization of proteins involved in clathrin assembly. J Cell Biol 1988; 107:877-886.
21. Collins BM, McCoy AJ, Kent HM et al. Molecular architecture and functional model of the endocytic AP2 complex. Cell 2002; 109:523-535.
22. Owen DJ, Vallis Y, Noble ME et al. A structural explanation for the binding of multiple ligands by the α-adaptin appendage domain. Cell 1999; 97:805-815.
23. Traub LM, Downs MA, Westrich JL et al. Crystal structure of the α appendage of AP-2 reveals a recruitment platform for clathrin-coat assembly. Proc Natl Acad Sci USA 1999; 96:8907-8912.
24. Owen DJ, Vallis Y, Pearse BM et al. The structure and function of the β2-adaptin appendage domain. EMBO J 2000; 19:4216-4227.
25. Galluser A, Kirchhausen T. The β1 and the β2 subunits of the AP complexes are the clathrin coat assembly components. EMBO J 1993; 12:5237-5244.
26. ter Haar E, Harrison SC, Kirchhausen T. Peptide-in-groove interactions link target proteins to the beta-propeller of clathrin. Proc Natl Acad Sci USA 2000; 97:1096-1100.
27. Gaidarov I, Keen JH. Phosphoinositide-AP-2 interactions required for targeting to plasma membrane clathrin-coated pits. J Cell Biol 1999; 146:755-764.
28. Padron D, Wang YJ, Yamamoto M et al. Phosphatidylinositol phosphate 5-kinase Iβ recruits AP-2 to the plasma membrane and regulates rates of constitutive endocytosis. J Cell Biol 2003; 162:693-701.
29. Lundmark R, Carlsson SR. Regulated membrane recruitment of dynamin-2 mediated by sorting nexin 9. J Biol Chem 2004.
30. Praefcke GJ, McMahon HT. The dynamin superfamily: Universal membrane tubulation and fission molecules? Nat Rev Mol Cell Biol 2004; 5:133-147.
31. Soulet F, Yarar D, Leonard M et al. SNX9 regulates dynamin assembly and is required for efficient clathrin-mediated endocytosis. Mol Biol Cell 2005; 16, (in press).
32. Olusanya O, Andrews PD, Swedlow JR et al. Phosphorylation of threonine 156 of the μ2 subunit of the AP2 complex is essential for endocytosis in vitro and in vivo. Curr Biol 2001; 11:896-900.
33. Ricotta D, Conner SD, Schmid SL et al. Phosphorylation of the AP2 μ subunit by AAK1 mediates high affinity binding to membrane protein sorting signals. J Cell Biol 2002; 156:791-795.
34. Rohde G, Wenzel D, Haucke V. A phosphatidylinositol (4,5)-bisphosphate binding site within μ2-adaptin regulates clathrin-mediated endocytosis. J Cell Biol 2002; 158:209-214.
35. Jackson AP, Flett A, Smythe C et al. Clathrin promotes incopotration of cargo into coated pits by activation of the AP2 adaptor μ2 kinase. J Cell Biol 2003; 163:231-236.
36. Conner SD, Schroter T, Schmid SL. AAK1 mediated μ2 phosphorylation is stimulated by assembled clathrin. Traffic 2003; 4:885-890.
37. Nesterov A, Carter RE, Sorkina T et al. Inhibition of the receptor-binding function of clathrin adaptor protein AP-2 by dominant-negative mutant μ2 subunit and its effects on endocytosis. EMBO J 1999; 18:2489-2499.
38. Ehrlich M, Boll W, Van Oijen A et al. Endocytosis by random initiation and stabilization of clathrin-coated pits. Cell 2004; 118:591-605.
39. Gonzalez-Gaitan M, Jackle H. Role of Drosophila α-adaptin in presynaptic vesicle recycling. Cell 1997; 88:767-776.
40. Shim J, Lee J. Molecular genetic analysis of apm-2 and aps-2, genes encoding the medium and small chains of the AP-2 clathrin-associated protein complex in the nematode Caenorhabditis elegans. Mol Cells 2000; 10:309-316.
41. Huang KM, D'Hondt K, Riezman H et al. Clathrin functions in the absence of heterotetrameric adaptors and AP180-related proteins in yeast. EMBO J 1999; 18:3897-3908.
42. Yeung BG, Phan HL, Payne GS. Adaptor complex-independent clathrin function in yeast. Mol Biol Cell 1999; 10:3643-3659.
43. Hinrichsen L, Harborth J, Andrees L et al. Effect of clathrin heavy chain- and α-adaptin specific small interfering RNAs on endocytic accessory proteins and receptor trafficking in HeLa cells. J Biol Chem 2003; 278:45160-45170.
44. Motley A, Bright NA, Seaman MN et al. Clathrin-mediated endocytosis in AP-2-depleted cells. J Cell Biol 2003; 162:909-918.
45. Huang F, Khvorova A, Marshall W et al. Analysis of clathrin-mediated endocytosis of epidermal growth factor receptor by RNA interference. J Biol Chem 2004; 279:16657-16661.
46. Robinson MS. Adaptable adaptors for coated vesicles. Trends Cell Biol 2004; 14:167-174.
47. Janvier K, Kato Y, Boehm M et al. Recognition of dileucine-based sorting signals from HIV-1 Nef and LIMP-II by the AP-1 γ-σ1 and AP-3 δ-σ3 hemicomplexes. J Cell Biol 2003; 163:1281-1290.
48. Marks MS, LW, Ohno H, Bonifacino JS. Protein targeting by tyrosine- and di-leucine-based signals: Evidence for distinct saturable components. J Cell Biol 1996; 135:341-354.

49. Carpentier JL, Gorden P, Anderson RG et al. Colocalization of[125] I-epidermal growth factor and ferritin-low density lipoprotein in coated pits: A quantitative electron microscopic study in normal and mutant human fibroblasts. J Cell Biol 1982; 95:73-77.
50. Morris SM, Cooper JA. Disabled-2 colocalizes with the LDLR in clathrin-coated pits and interacts with AP-2. Traffic 2001; 2:111-123.
51. Mishra SK, Keyel PA, Hawryluk MJ et al. Disabled-2 exhibits the properties of a cargo-selective endocytic clathrin adaptor. EMBO J 2002; 21:4915-4926.
52. Stang E, Blystad FD, Kazazic M et al. Cbl-dependent ubiquitination is required for progression of EGF receptors into clathrin-coated pits. Mol Biol Cell 2004; 15:3591-3604.
53. Sorkin A, McKinsey T, Shih W et al. Stoichiometric interaction of the epidermal growth factor receptor with the clathrin-associated protein complex AP-2. J Biol Chem 1995; 270:619-625.
54. Nesterov A, Wiley HS, Gill GN. Ligand-induced endocytosis of epidermal growth factor receptors that are defective in binding adaptor proteins. Proc Natl Acad Sci USA 1995; 92:8719-8723.
55. Warren RA, Green FA, Enns CA. Saturation of the endocytic pathway for the transferrin receptor does not affect the endocytosis of the epidermal growth factor receptor. J Biol Chem 1997; 272:2116-2121.
56. Warren RA, Green FA, Stenberg PE et al. Distinct saturable pathways for the endocytosis of different tyrosine motifs. J Biol Chem 1998; 273:17056-17063.
57. Maxfield FR, Schlessinger J, Shechter Y et al. Collection of insulin, EGF and α2-macroglobulin in the same patches on the surface of cultured fibroblasts and common internalization. Cell 1978; 14:805-810.
58. Ajioka RS, Kaplan J. Characterization of endocytic compartments using the horseradish peroxidase-diaminobenzidine density shift technique. J Cell Biol 1987; 104:77-85.
59. Sorkin A, Von Zastrow M. Signal transduction and endocytosis: Close encounters of many kinds. Nat Rev Mol Cell Biol 2002; 3:600-614.
60. Marchese A, Chen C, Kim YM et al. The ins and outs of G protein-coupled receptor trafficking. Trends Biochem Sci 2003; 28:369-376.
61. Lefkowitz RJ, Whalen EJ. β-arrestins: Traffic cops of cell signaling. Curr Opin Cell Biol 2004; 16:162-168.
62. Gaidarov I, Krupnick JG, Falck JR et al. Arrestin function in G protein-coupled receptor endocytosis requires phosphoinositide binding. EMBO J 1999; 18:871-881.
63. Laporte SA, Oakley RH, Holt JA et al. The interaction of β-arrestin with the AP-2 adaptor is required for the clustering of β2-adrenergic receptor into clathrin-coated pits. J Biol Chem 2000; 275:23120-23126.
64. Milano SK, Pace HC, Kim YM et al. Scaffolding functions of arrestin-2 revealed by crystal structure and mutagenesis. Biochemistry 2002; 41:3321-3328.
65. Goodman Jr OB, Krupnick JG, Santini F et al. β-arrestin acts as a clathrin adaptor in endocytosis of the β2- adrenergic receptor. Nature 1996; 383:447-450.
66. Krupnick JG, Goodman Jr OB, Keen Jr JH. Arrestin/clathrin interaction. Localization of the clathrin binding domain of nonvisual arrestins to the carboxy terminus. J Biol Chem 1997; 272:15011-15016.
67. Santini F, Penn RB, Gagnon AW et al. Selective recruitment of arrestin-3 to clathrin coated pits upon stimulation of G protein-coupled receptors. J Cell Sci 2000; 113(Pt 13):2463-2470.
68. Chen W, Ren XR, Nelson CD et al. Activity-dependent internalization of smoothened mediated by β2 and GRK2. Science 2004; 306:2257-2260.
69. Chen W, ten Berge D, Brown J et al. Dishevelled 2 recruits β-arrestin 2 to mediate Wnt5A-stimulated endocytosis of Frizzled 4. Science 2003; 301:1391-1394.
70. Chen W, Kirkbride KC, How T et al. β-arrestin 2 mediates endocytosis of type III TGF-β receptor and down-regulation of its signaling. Science 2003; 301:1394-1397.
71. Wu JH, Peppel K, Nelson CD et al. The adaptor protein β-arrestin2 enhances endocytosis of the low-density lipoprotein receptor. J Biol Chem 2003.
72. Slepnev VI, De Camilli P. Accessory factors in clathrin-dependent synaptic vesicle endocytosis. Nat Rev Neurosci 2000; 1:161-172.
73. McMahon HT, Mills IG. COP and clathrin-coated vesicle budding: Different pathways, common approaches. Curr Opin Cell Biol 2004; 16:379-391.
74. Sorkin A. Cargo recognition during clathrin-mediated endocytosis: A team effort. Curr Opin Cell Biol 2004; 16:392-399.
75. Blaikie P, Immanuel D, Wu J et al. A region in Shc distinct from the SH2 domain can bind tyrosine-phosphorylated growth factor receptors. J Biol Chem 1994; 269:32031-32034.
76. Yan KS, Kuti M, Zhou MM. PTB or not PTB — that is the question. FEBS Lett 2002; 513:67-70.
77. Davis CG, Lehrman MA, Russell DW et al. The J.D. mutation in familial hypercholesterolemia: Amino acid substitution in cytoplasmic domain impedes internalization of LDL receptors. Cell 1986; 45:15-24.

78. Garcia CK, Wilund K, Arca M et al. Autosomal recessive hypercholesterolemia caused by muta-
    tions in a putative LDL receptor adaptor protein. Science 2001; 292:1394-1398.
79. Morris SM, Tallquist MD, Rock CO et al. Dual roles for the Dab2 adaptor protein in embryonic
    development and kidney transport. EMBO J 2002; 21:1555-1564.
80. Jones C, Hammer RE, Li WP et al. Normal sorting, but defective endocytosis of the LDL receptor
    in mice with autosomal recessive hypercholesterolemia. J Biol Chem 2003; 278:29024-29030.
81. Tietge UJ, Genschel J, Schmidt HH. A Q136Stop mutation in the ARH gene causing autosomal
    recessive hypercholesterolaemia with severely delayed LDL catabolism. J Intern Med 2003;
    253:582-583.
82. Eden ER, Patel DD, Sun X et al. Restoration of LDL-receptor function in cells from patients with
    autosomal recessive hypercholesterolemia by retroviral expression of ARH1. J Clin Invest 2002;
    110:1695-1702.
83. Al-Kateb H, Bahring S, Hoffmann K et al. Mutation in the ARH gene and a chromosome 13q
    locus influence cholesterol levels in a new form of digenic-recessive familial hypercholesterolemia.
    Circ Res 2002; 90:951-958.
84. Canizales-Quinteros S, Aguilar-Salinas CA, Huertas-Vazquez A et al. A novel ARH splice site
    mutation in a Mexican kindred with autosomal recessive hypercholesterolemia. Hum Genet 2005;
    116:114-120.
85. Harada-Shiba M, Takagi A, Miyamoto Y et al. Clinical features and genetic analysis of autosomal
    recessive hypercholesterolemia. J Clin Endocrinol Metab 2003; 88:2541-2547.
86. Barbagallo CM, Emmanuele G, Cefalu AB et al. Autosomal recessive hypercholesterolemia in a
    Sicilian kindred harboring the 432insA mutation of the ARH gene. Atherosclerosis 2003;
    166:395-400.
87. Osono Y, Woollett LA, Herz J et al. Role of the low density lipoprotein receptor in the flux of
    cholesterol through the plasma and across the tissues of the mouse. J Clin Invest 1995;
    95:1124-1132.
88. Nagai J, Christensen EI, Morris SM et al. Mutually-dependent localization of megalin and Dab2
    in the renal proximal tubule. Am J Physiol Renal Physiol 2005:(in press).
89. Nykjaer A, Dragun D, Walther D et al. An endocytic pathway essential for renal uptake and
    activation of the steroid 25-(OH) vitamin D3. Cell 1999; 96:507-515.
90. Mishra SK, Watkins SC, Traub LM. The autosomal recessive hypercholesterolemia (ARH) protein
    interfaces directly with the clathrin-coat machinery. Proc Natl Acad Sci USA 2002; 99:16099-16104.
91. Stolt PC, Jeon H, Song HK et al. Origins of peptide selectivity and phosphoinositide binding
    revealed by structures of Disabled-1 PTB domain complexes. Structure (Camb) 2003; 11:569-579.
92. Yun M, Keshvara L, Park CG et al. Crystal structures of the dab homology domains of mouse
    disabled 1 and 2. J Biol Chem 2003; 278:36572-36581.
93. Morris AJ, Frohman MA, Engebrecht J. Measurement of phospholipase D activity. Anal Biochem
    1997; 252:1-9.
94. He G, Gupta S, Yi M et al. ARH is a modular adaptor protein that interacts with the LDL
    receptor, clathrin and AP-2. J Biol Chem 2002; 277:44044-44049.
95. Brett TJ, Traub LM, Fremont DH. Accessory protein recruitment motifs in clathrin-mediated en-
    docytosis. Structure (Camb) 2002; 10:797-809.
96. Miele AE, Watson PJ, Evans PR et al. Two distinct interaction motifs in amphiphysin bind two
    independent sites on the clathrin terminal domain β-propeller. Nat Struct Mol Biol 2004;
    11:242-248.
97. Praefcke GJ, Ford MG, Schmid EM et al. Evolving nature of the AP2 α-appendage hub during
    clathrin-coated vesicle endocytosis. EMBO J 2004; 23:4371-4383.
98. Owen DJ. Linking endocytic cargo to clathrin: Structural and functional insights into coated vesicle
    formation. Biochem Soc Trans 2004; 32:1-14.
99. Shen Q, Temple S. Creating asymmetric cell divisions by skewing endocytosis. Sci STKE 2002;
    2002:PE52.
100. Rhyu MS, Jan LY, Jan YN. Asymmetric distribution of numb protein during division of the sen-
    sory organ precursor cell confers distinct fates to daughter cells. Cell 1994; 76:477-491.
101. O'Connor-Giles KM, Skeath JB. Numb inhibits membrane localization of Sanpodo, a four-pass
    transmembrane protein, to promote asymmetric divisions in Drosophila. Dev Cell 2003; 5:231-243.
102. Wendland B. Epsins: Adaptors in endocytosis? Nat Rev Mol Cell Biol 2002; 3:971-977.
103. Hicke L, Dunn R. Regulation of membrane protein transport by ubiquitin and ubiquitin-binding
    proteins. Annu Rev Cel Dev Biol 2003; 19:141-172.
104. Ford MG, Mills IG, Peter BJ et al. Curvature of clathrin-coated pits driven by epsin. Nature
    2002; 419:361-366.

105. Hofmann K, Falquet L. A ubiquitin-interacting motif conserved in components of the proteasomal and lysosomal protein degradation systems. Trends Biochem Sci 2001; 26:347-350.
106. Chen H, Fre S, Slepnev VI et al. Epsin is an EH-domain-binding protein implicated in clathrin-mediated endocytosis. Nature 1998; 394:793-797.
107. Drake MT, Downs MA, Traub LM. Epsin binds to clathrin by associating directly with the clathrin-terminal domain: Evidence for cooperative binding through two discrete sites. J Biol Chem 2000; 275:6479-6489.
108. Rosenthal JA, Chen H, Slepnev VI et al. The epsins define a family of proteins that interact with components of the clathrin coat and contain a new protein module. J Biol Chem 1999; 274:33959-33965.
109. Tian X, Hansen D, Schedl T et al. Epsin potentiates Notch pathway activity in Drosophila and C. elegans. Development 2004; 131:5807-5815.
110. Overstreet E, Fitch E, Fischer JA. Fat facets and Liquid facets promote Delta endocytosis and Delta signaling in the signaling cells. Development 2004; 131:5355-5366.
111. Lai EC, Roegiers F, Qin X et al. The ubiquitin ligase Drosophila Mind bomb promotes Notch signaling by regulating the localization and activity of Serrate and Delta. Development 2005; 132:2319-2332.
112. Wang W, Struhl G. Drosophila Epsin mediates a select endocytic pathway that DSL ligands must enter to activate Notch. Development 2004; 131:5367-5380.
113. Le Borgne R, Bardin A, Schweisguth F. The roles of receptor and ligand endocytosis in regulating Notch signaling. Development 2005; 132:1751-1762.
114. Chen X, Zhang B, Fischer JA. A specific protein substrate for a deubiquitinating enzyme: Liquid facets is the substrate of Fat facets. Genes Dev 2002; 16:289-294.
115. Metzler M, Li B, Gan L et al. Disruption of the endocytic protein HIP1 results in neurological deficits and decreased AMPA receptor trafficking. EMBO J 2003; 22:3254-3266.
116. Nonet ML, Holgado AM, Brewer F et al. UNC-11, a Caenorhabditis elegans AP180 homologue, regulates the size and protein composition of synaptic vesicles. Mol Biol Cell 1999; 10:2343-2360.
117. Mossessova E, Bickford LC, Goldberg J. SNARE selectivity of the COPII coat. Cell 2003; 114:483-495.
118. Traub LM. Common principles in clathrin-mediated sorting at the Golgi and the plasma membrane. Biochim Biophys Acta 2005; in press.
119. Merrifield CJ. Seeing is believing: Imaging actin dynamics at single sites of endocytosis. Trends Cell Biol 2004; 14:352-358.
120. Blagoev B, Ong SE, Kratchmarova I et al. Temporal analysis of phosphotyrosine-dependent signaling networks by quantitative proteomics. Nat Biotechnol 2004; 22:1139-1145.
121. Barabasi AL, Oltvai ZN. Network biology: Understanding the cell's functional organization. Nat Rev Genet 2004; 5:101-113.

CHAPTER 7

Protein Sorting in Endosomes

Krupa Pattni and Harald Stenmark*

Abstract

Molecules delivered to endosomes by endocytosis or biosynthetic trafficking can be either recycled to the cell surface, transported to lysosomes, or shunted retrogradely to the biosynthetic pathway. The distinct fates of different endosomal cargo molecules point to the existence of sorting machineries able to distinguish between cargoes. In this review we will highlight recent studies that are beginning to elucidate the endosomal sorting machineries that recognize different cargoes, as well as individual sorting signals that specify their destinations.

Introduction

Endocytosed or biosynthesized molecules that transit through endosomes have several possible destinations (see Fig. 1). They can be either sorted for recycling to the plasma membrane, anterograde trafficking to the degradative lysosomes or retrograde transport to the trans-Golgi network of the biosynthetic pathway. Since different endosomal cargo molecules have distinct itineraries, efficient sorting mechanisms must exist that recognize specific cargoes. Accumulating evidence suggests that all the above-mentioned trafficking routes out of endosomes rely on the recognition of specific cargo sorting determinants by distinct endosomal sorting machineries. Here, we will discuss emerging data that are beginning to shed light on the sorting determinants of endosomal cargo proteins and the machineries that recognize them.

Endosomes As Sorting Stations in Intracellular Membrane Trafficking

The organisation of the endocytic pathway is reviewed elsewhere in this book (Chapter 1). For the purpose of this review, we will distinguish between early endosomes (EEs), recycling endosomes (REs) and late endosomes (LEs). As outlined in Figure 1, recycling to the plasma membrane can either occur directly from the EEs or indirectly via the RE, in processes controlled by the small GTPases Rab4 and Rab11, respectively.[1] Typical examples of recycled membrane proteins include the receptors for transferrin and low-density lipoprotein.[2] In EEs, sorting towards the degradative pathway also takes place, as exemplified by ligand-activated growth factor receptors such as the epidermal growth factor (EGF) receptor.[3] Most membrane proteins destined for LEs and lysosomes are targeted into intraluminal vesicles that invaginate from the limiting membrane of the EE. Sorting to the trans-Golgi network (TGN) can occur from both EEs and LEs, most probably through different sorting machineries (see below).

The architecture of various types of endosomes reflects their specific purposes. For instance, the tubular morphology of REs and the "recycling" part of EEs ensures a high

*Corresponding Author: Harald Stenmark—Department of Biochemistry, The Norwegian Radium Hospital, Montebello, N-0310 Oslo, Norway. Email: stenmark@ulrik.uio.no

Endosomes, edited by Ivan Dikic. ©2006 Landes Bioscience and Springer Science+Business Media.

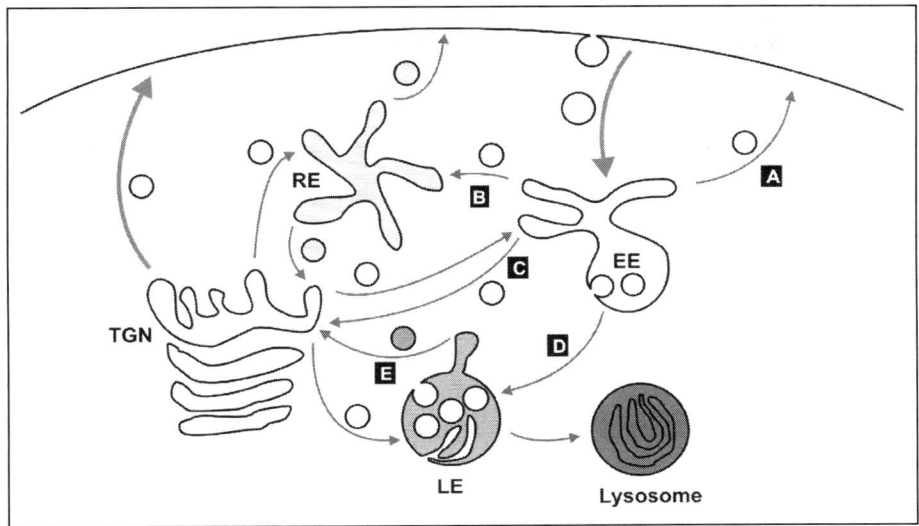

Figure 1. Protein sorting in the endocytic pathway of a nonpolarized cell. Upon endocytosis, cargo is transferred to the early endosome (EE). From here, cargo can be sorted for direct recycling to the plasma membrane (A), recycling via the recycling endosome (RE) (B), transport to the trans-Golgi network (TGN) (C) or transport to the late endosome (LE) (D). Sorting to the TGN can alternatively occur from the LE (E). Various trafficking steps are indicated by arrows whose line thickness reflects relative importance.

membrane-to-volume ratio. This favours an enrichment, and thus sorting, of endosomal membrane proteins with respect to soluble content, into transport carriers that leave the endosomal tubules.[2] In this way, small tubules and vesicles that bud from the REs and the tubular regions of the EEs are efficient vehicles for membrane proteins, mostly targeted for recycling to the plasma membrane. Another peculiar geometric feature of endosomes is the invagination of the cisternal part of the endosome membrane to form intraluminal vesicles (see Fig. 1). Since the membrane of such vesicles is more accessible to digestion by lysosomal enzymes than the limiting membrane of LEs (probably due to differences in lipid compositions and lower abundance of highly glycosylated membrane proteins), such vesicles are ideally suited as vehicles for membrane proteins destined for degradation.[4] The molecular machineries responsible for the formation of intraluminal vesicles, and for the sorting of membrane proteins into them, are beginning to emerge, and this will be discussed in the following sections (Table 1).

Sorting to the Recycling Route

The transferrin receptor (TfR) has served as a prototypic example of a recycling membrane protein. This receptor is constitutively endocytosed from clathrin-coated pits regardless of ligand binding. Upon reaching the EE, the TfR is very efficiently recycled to the plasma membrane. This recycling occurs both directly and via the RE.[5] The finding that a truncated TfR lacking the whole cytoplasmic tail recycles at the same rate as the wild-type receptor[6] initially led to the conclusion that endocytic recycling is signal independent.[7] However, more recent studies have revealed that multiple receptors, including the TfR, contain bona fide recycling determinants. The TfR has been found to contain two phenylalanine-based signals which, when mutated to alanine, slow down TfR trafficking from and to the RE.[8] These recycling sorting signals interact with ACAP1, a GTPase-activating protein for Arf6, which promotes cargo sorting to enhance TfR recycling.[8] In addition, ACAP1 interacts with cellubrevin, another recycling cargo

Table 1. Protein sorting in endosomes

Sorting Step (See Fig. 1)	Cargo (Example)	Sorting Signal	Sorting Machinery	Refs.
EE → PM (A)	TfR	?	Rab4, AP-1, Hrs, actinin-4, BERP, myosin-V	9,82
	β2-AR	Ser-411 phosphorylation by GRK5	EPB50, Hrs	12,13
	μ-opioid receptor	LENLEAE motif	Hrs	13
	D1 dopamine receptor	25-aa C-terminal sequence	?	16
	LHR	Palmitoylation	?	83
EE → RE → PM (B)	TfR	LF and RF motifs	Rab11, ACAP1	8
	Cellubrevin	?	Rab11, ACAP1	8
EE → TGN (C)	CI-M6PR	?	Retromer, EpsinR	61,68
	TGN38	SXYQRL motif	EpsinR	68,84
	Shiga toxin	(The receptor for Shiga toxin is the lipid Gb3)	Clathrin, EpsinR	68,69
EE → LE (D)	activated EGFR	Ubiquitination	Hrs, STAM, ESCRTs, GGA3	17,40,41, 85-87
	activated CXCR4	Ubiquitination	Hrs	79
	misfolded CFTR	Ubiquitination	Hrs, ESCRTs	88
	δ-opioid receptor	C-terminal α-helical region of cytosolic tail	GASP	52
	LAMP-1	C-terminal YQTI motif	AP-3	55,89
LE → TGN (E)	CD-M6PR	LL and FW motifs	Rab9, TIP47	90-92
	CI-M6PR	PPAPRPG motif	Rab9, TIP47	90,91
	Furin	SDSEED motif	?	93

The table refers to the sorting steps indicated in Figure 1 and provides examples of well-studied cargo molecules in mammalian cells as well as, when known, the signals and machineries that mediate their sorting. Note that some cargoes, including TfR and M6PR, can be sorted along at least two distinct pathways, and that the distinctions between pathways A and B have so far been incompletely defined for most cargoes. β2-AR: β2-adrenergic receptor; LHR: luteinizing hormone receptor; LAMP: lysosome-associated membrane protein; CD: cation-dependent; CI: cation-independent; CFTR: cystic fibrosis transmembrane conductance regulator; CXCR4: a chemochine receptor.

protein, but not with Lamp1, a cargo protein transported to lysosomes. ACAP1 may thus have a general function in endocytic recycling via the RE.

A recent study has shed light on how the TfR may be differentially recycled directly to the plasma membrane. A complex called CART (cytoskeleton-associated recycling or transport), consisting of hepatocyte growth factor regulated tyrosine kinase substrate (Hrs), actinin-4, brain-expressed RING-finger protein (BERP) and myosin-V, was found essential for fast recycling of TfRs. Disruption of this complex led to a slower recycling via the RE.[9] Thus, it is likely that ACAP1 sorts the TfR into the indirect recyling pathway.

How can these findings be reconciled with the efficient recycling of the truncated TfR?[6] One possibility is that the cytoplasmic tail of the TfR may contain endosomal retention signals in addition to recycling signals. Such signals, one consisting of an acidic cluster and one consisting of a di-leucine motif, have already been identified in insulin-regulated aminopeptidase, a slowly recycling membrane protein.[10] Even though these retention signals are reminiscent of endocytic signals recognized by adaptor protein complexes, the specific machinery that recognises the endocytic retention signals has not yet been identified.

The first demonstrations of endocytic recycling signals have come from studies of G-protein-coupled receptors (GPCRs).[11] Most GPCRs are internalised in response to ligand binding, and many GPCRs display rapid recycling. GRK-5-mediated phosphorylation of serine-411 in the cytoplasmic tail of the β2-adrenergic receptor is required for efficient recycling of this GPCR. Sorting into the recycling pathway appears to be mediated by an interaction of the phosphorylated receptor with the PDZ domain of EBP50 (ezrin-radixin-moesin-binding phosphoprotein 50), a protein associated with the actin cytoskeleton.[12] Hrs, a constituent of the CART complex that mediates rapid recycling of TfRs, has recently been found essential for efficient recycling of the β2-adrenergic receptor as well as for recycling of another GPCR, the μ-opioid receptor. The N-terminal VHS-domain seems crucial for this atypical function of Hrs.[13] So far, the functional relationship between Hrs/CART and EBP50 in GPCR sorting has not been clarified.

While the β2-adrenergic and μ-opioid receptors appear to employ related mechanisms for their sequence-dependent recycling, other GPCRs probably use distinct mechanisms for their sorting into the recycling pathway. The lutropin receptor contains a 17-residue membrane-distal cytosolic signal that is necessary and sufficient for recycling of the endocytosed receptor.[14,15] In this receptor, specific leucine, cysteine, glycine and threonine residues have been implicated in the recycling function. Likewise, the D1 dopamine receptor contains a distinct 23-residue membrane-proximal sequence that mediates its recycling.[16] All the above-mentioned short sequences can be transplanted onto nonrecycling model receptors and mediate their recycling, suggesting that they interact directly with the sorting machineries that mediate endocytic recycling. So far, these machineries are not known, but the tools should now be available for their identification.

Sorting to the Degradative Pathway

In mammalian cells, ligand-bound (activated) growth factor receptors serve as prototypic examples of membrane proteins that are delivered to the degradative pathway after endocytosis (see chapter 9). As discussed in the following section, covalent attachment of mono-ubiquitin to one or several lysine residues in the cytoplasmic regions of a membrane protein represents an important sorting signal for the degradative pathway.[17,18]

Mono-Ubiquitin As a Degradative Sorting Signal

A number of membrane proteins are mono-ubiquitinated at the plasma membrane or on endosomes. This posttranslational modification has been shown to serve as an endocytosis signal in some cases, but the most general function of mono-ubiquitination in membrane traffic appears to be its role in endosomal sorting.[19] In fact, mono-ubiquitin serves as a dominant signal for degradative protein sorting, as illustrated by the fact that recombinant ubiquitin fusions of endocytic membrane proteins are efficiently targeted to lysosomes.[20,21] Even though a single mono-ubiquitin moiety may be sufficient for degradative protein sorting,[17,19,22] many membrane proteins, including growth factor receptors, are mono-ubiquitinated at multiple lysine residues (multi-ubiquitinated; see Fig. 2).[17,18] Mono- and multi-ubiquitination are mediated by the same sets of substrate-specific ubiquitin ligases, whose impaired function may be associated with diseases such as cancer (see Chapter 9). Recent work has shed light on the mechanisms of sorting of ubiquitinated membrane proteins. It is now clear that ubiquitinated cargo is recognised by a number of sorting components that contain specialised ubiquitin-binding domains (UBDs; see Fig. 2).[23]

Eight UBDs have been identified to date. These include the UIM (ubiquitin-interacting motif),[24] UBA (ubiquitin-associated domain),[25] UBC (ubiquitin-conjugating enzyme-like)/ UEV (ubiquitin E2 variant),[26,27] CUE (Cue1-homologous),[28,29] GAT (GGA and TOM1),[32] GLUE,[33] PAZ (poly-ubiquitin-associated zinc finger),[30] and NZF (novel zinc finger)[31]

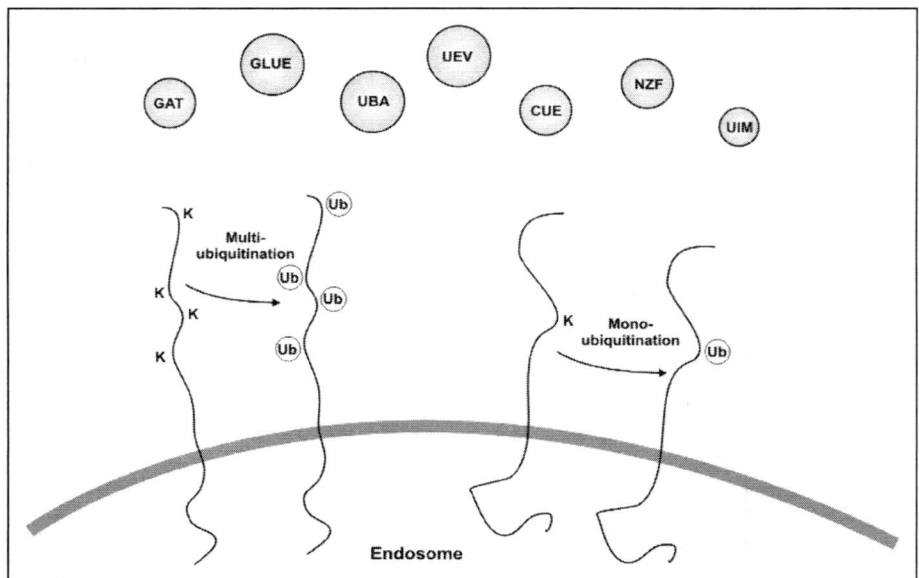

Figure 2. Mono- and multi-ubiquitin as sorting signals. Mono- and multi-ubiquitinations of endosomal membrane proteins, catalysed by substrate-specific E3 ubiquitin ligases, are illustrated schematically.[23] The mono- and multi-ubiquitinated proteins are recognised by the various UBDs indicated on top. The avidity for multi-ubiquitinated cargo is probably increased by multimerisation of UBD-containing proteins.

domains. With the exception of the PAZ domain, all the UBDs have been found within various subunits of endosomal sorting complexes (see Table 2). Even though these UBDs have diverse structural folds, they all contact the hydrophobic Ile-44 patch of ubiquitin. The affinity of this interaction is rather weak and is stronger for poly-ubiquitin chains than for mono-ubiquitin.[19] The affinity and specificity of the interaction to mono-ubiquitin is probably increased in vivo by additional interactions with the modified protein or formation of multimeric complexes of ubiquitin-binding proteins.[17]

Ubiquitin-Dependent Degradative Protein Sorting

Distinct UBD-containing proteins complexes are recruited to the plasma membrane and to endosomes. In the endosome membrane, the ubiquitinated membrane protein is recognised by a UBD-containing machinery that sorts it into intraluminal vesicles. The ubiquitin-binding endosomal protein Hrs is central to this machinery.[34] This protein is recruited to endosomal membranes through binding of its FYVE domain[35] to phosphatidylinositol 3-phosphate (PI3P), a phosphoinositide localized specifically to these membranes.[36,37] The C-terminus of Hrs binds clathrin,[38] and Hrs is found in a characteristic flat clathrin coat on EEs.[21,39] Hrs recognises ubiquitinated cargo via its UIM.[21] Although Hrs binds ubiquitin with low affinity, its avidity for multi-ubituitinated cargo may be increased by the fact that membrane-bound Hrs is complexed with two other ubiquitin-binding proteins, signal-transducing adaptor molecule (STAM) and Eps15 (see Table 2).[40] The function of the clathrin lattice could be to concentrate Hrs/STAM/Eps15 in restricted microdomains in order to increase sorting efficacy. Interestingly, another clathrin-binding protein, GGA3 (Golgi-localizing, gamma-adaptin ear domain homology, ARF-binding protein 3), has recently been found to interact with and sort ubiquitinated cargo for degradation.[41] The relationship between GGA3 and Hrs will thus be

Table 2. Ubiquitin-binding proteins involved in degradative endosomal sorting

Protein	Organism	Function	UBD	References
Rabex-5	Mammals	GEF for the GTPase Rab5, a regulator of trafficking through early endosomes	?	94
Vps9	Yeast	Homologue of Rabex-5	CUE	28,95
Eps15	Mammals	Endocytosis, endosomal sorting, in complex with Hrs and STAM on endosomes	UIM	40,96
Ede1	Yeast	Homologue of Eps15	UBA	97
GGA3	Mammals	TGN-/endosomal sorting, interacts with Tsg101	GAT	41,98
Gga1/Gga2	Yeast	Homologues of GGA3	GAT	99
Tom1	Mammals	Endosomal sorting, in complex with Tollip and Endofin	GAT	32,100
Tom1L1	Mammals	Endosomal sorting, interacts with Hrs and Tsg101	GAT	101
Hrs	Mammals	Endosomal sorting, in complex with STAM and Eps15, recruits ESCRT-I	UIM	21,40,45, 86,102,103
Vps27	Yeast	Homologue of Hrs	UIM	97,104
STAM	Mammals	Endosomal sorting, in complex with Hrs and Eps15	UIM,VHS*	40,105,106
Hse1	Yeast	Homologue of STAM	UIM	104,107
Tsg101	Mammals	Endosomal sorting, subunit of ESCRT-I	UEV	47,86
Vps23	Yeast	Homologue of Tsg101	UEV	26,49
Eap45	Mammals	Endosomal sorting, subunit of ESCRT-II	GLUE	33
Vps36	Yeast	Homologue of Eap45	NZF	43,108

Endosomal UBD-containing proteins from yeast and mammals are listed. UBD-containing proteins thought to function mainly in the TGN or at the plasma membrane have not been included. *In the case of mammalian STAM, a VHS (Vps27, Hrs, STAM) domain has been implicated in ubiquitin-binding in addition to the UIM.

interesting to examine. Hrs has a function that is conserved from yeast to man, and the sorting machinery immediately downstream of Hrs is also highly conserved.[42] Three endosomal sorting complexes required for transport (ESCRTs) were first identified in yeast and subsequently shown to be conserved in mammals.[26,43,44] Current evidence suggests that ESCRT-I is recruited to endosome membranes by a direct interaction with Hrs,[45,46] that ESCRT-II is recruited by binding to ESCRT-I,[43] and that ESCRT-III is recruited by ESCRT-II.[44] Both ESCRT-I and ESCRT-II contain ubiquitin-binding subunits (see Table 2), and it is conceivable that ubiquitinated cargo is transferred from Hrs to ESCRT-II via ESCRT-I (Fig. 3). Whereas ESCRT-I and ESCRT-II have defined biochemical compositions, and their partial crystal structures have been solved,[47-50] the composition of ESCRT-III is less well understood. This complex consists of two heterodimeric subcomplexes that assemble into large multimers on membranes (Fig. 3). It is thought that ESCRT-III multimerisation contributes to form inward invaginations and vesiculation of the endosome membrane, by a hitherto undefined mechanism.[44] The AAA-ATPase Vps4 mediates the disassembly of ESCRT-III complexes, thus allowing the same complexes to participate in multiple rounds of transport.[51]

Ubiquitin-Independent Degradative Sorting

Even though ubiquitin serves as a widespread sorting signal for the degradative pathway, there are also examples of nonubiquitinated membrane proteins that reach this pathway.[20,52]

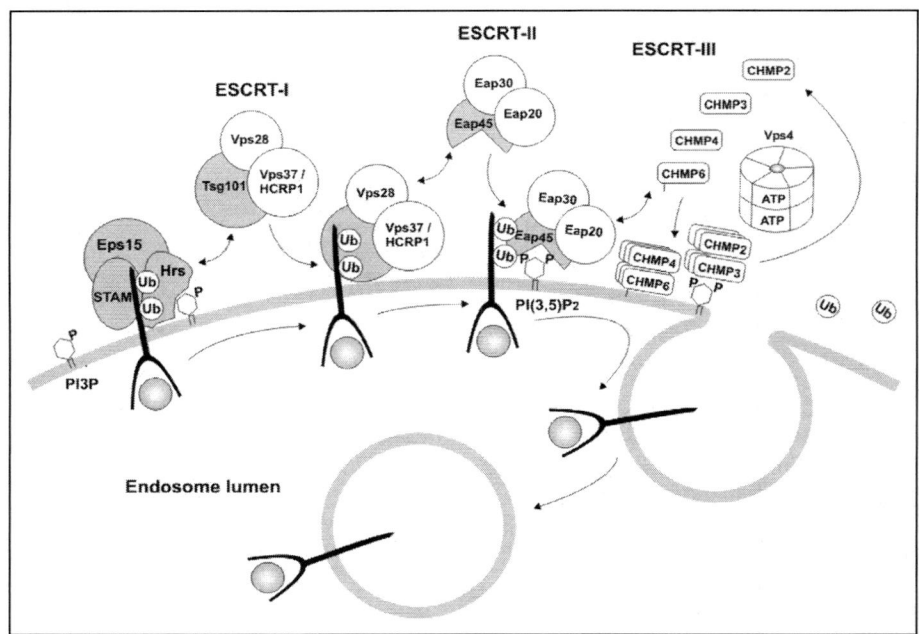

Figure 3. Ubiquitin-mediated sorting into the degradative pathway A ubiquitinated membrane protein in the endosome membrane is recognised by a Hrs-containing complex. The cargo is delivered first to ESCRT-I, thereafter to ESCRT-II and finally transported into intraluminal invaginations via the function of ESCRT-III. Disassembly of ESCRT-III is mediated by the ATPase Vps4.[51,109] The phosphoinositides phosphatidylinositol 3-phosphate (PI3P) and phosphatidylinositol 3,5-bisphosphate [PI(3,5)P$_2$] participate in the membrane recruitment of Hrs and ESCRTs, respectively.[33,37,110] Ubiquitin-binding proteins are shaded in gray. Double arrows indicate interactions between various complexes. Ub, ubiquitin.

So far, it is not known whether such cargo requires its own sorting machinery, whether it can piggy-back on ubiquitinated cargo, or whether the Hrs/ESCRT-containing sorting machinery can directly recognise even nonubiquitinated cargo. A GPCR, the δ-opioid receptor, is targeted for lysosomal degradation by a mechanism that does not require ubiquitination of the receptor. Interestingly, the lysosomal targeting of this receptor appears to require Hrs but not ESCRT-I.[53] Hrs is thus an extremely versatile sorting receptor since it is used for endocytic recycling of TfRs, β2-adrenergic receptors and μ-opioid receptors, degradative sorting of ubiquitinated cargoes, as well as for degradative sorting of nonubiquinated cargoes (see Table 1). The various sorting functions of Hrs are probably mediated by distinct domains. In addition to Hrs, the ubiquitin-independent endosomal sorting machinery is likely to use some unique components not found in the ubiquitin-dependent cargo-sorting machinery. For example, the GPCR associated sorting protein (GASP) binds to the cytoplasmic tail of the δ-opioid receptor, and this interaction appears necessary for the lysosomal targeting of the receptor.[54] However, so far it is not known whether GASP and Hrs operate in the same sorting pathway of nonubiquitinated cargo.

Sorting from EEs to LEs does not exclusively occur via inclusion into intraluminal vesicles. An alternative pathway, followed by resident lysosomal membrane proteins such as LAMP-1, appears to involve tubular extensions of the EE.[55] Even though such extensions have typically been associated with endocytic recycling, recent evidence suggests that LAMP-1 is enriched in specific endosomal tubules that contain the adaptor complex AP-3. Moreover, knockdown of AP-3 strongly increases recycling of LAMP-1 at the expense of its transport to lysosomes.[55] This suggests the existence of a specialized AP-3- and tubule-driven sorting pathway for resident lysosomal membrane proteins.

Sorting to the Biosynthetic Pathway

Like trafficking between the plasma membrane and endosomes, trafficking between the biosynthetic and endocytic pathways is bidirectional. The mannose 6-phosphate receptor (M6PR) is a well-studied example of a membrane protein that shuttles between the biosynthetic and endocytic pathways.[56] Upon delivery of newly synthesised lysosomal hydrolases to the acidic lumen of endosomes, the M6PR is recycled from these organelles to the TGN where it can capture new cargo. The sorting of M6PR in LEs is mediated by the tail-interacting protein of 47 kDa (TIP47), which binds to the cytoplasmic tail of the M6PR.[57] This binding is cooperative with a binding of TIP47 to the active (GTP-bound) form of Rab9, which controls vesicle trafficking between LEs and the TGN, suggesting a coupling between the budding of MP6R-containing vesicles and the function of Rab9.[58] According to the current models of endosomal protein sorting, the localisation of M6PR within LE membranes is somewhat surprising. The receptor can be detected on intraluminal invaginations or vesicles,[59] structures that are reminiscent to those involved in sorting to the degradative pathway.[3,42] This indicates that distinct subpopulations of intraluminal invaginations/vesicles exist, and that intraluminal membranes are not necessarily destined for degradation (see chapter 2 for a detailed discussion about the dynamics of intraluminal vesicles). The biogenesis of the carriers for trafficking between LEs and the TGN remains to be clarified.

M6PRs are not only recycled from LEs to the TGN - such transport can also occur from EEs. An evolutionarily conserved sorting complex, retromer,[60] is mainly localised to tubular extensions of EEs in mammalian cells.[61] This complex consists of the PI3P-binding sorting nexins 1 and 2 and the sorting components hVps26, hVps29 and hVps36,[62,63] of which the hVps35 component binds to the cytoplasmic tail of the M6PR. Depletion of retromer by RNA interference increases the lysosomal turnover of the M6PR, decreases cellular levels of lysosomal hydrolases, and causes swelling of lysosomes. These observations indicate that retromer prevents the delivery of the M6PR to lysosomes, probably by sequestration into endosome-derived tubules from where the receptor returns to the TGN.[61,64] The apparent lack of colocalization between retromer and Rab9/TIP47 suggests that these complexes are involved in M6PR retrieval at distinct locations. This raises the possibility that retromer and TIP47/Rab9 could function sequentially in endosome-to-TGN retrieval of M6PR as a mechanism to prevent mis-sorting of M6PRs to lysosomes. It is interesting to note that yeast Vps10, a sorting receptor for carboxypeptidase Y that bears structural and functional similarities to the mammalian M6PR, requires retromer for its trafficking from endosomes to the TGN.[65] Since yeast has no direct counterpart of Rab9, this suggests that the retromer has appeared earlier in evolution than Rab9/TIP47 for endosome to TGN sorting of M6PR-like molecules.

The trafficking of many bacterial and plant toxins that use cellular endocytotic and intracellular sorting machineries to invade cells has shed additional light on the understanding of mechanisms of endosomal sorting. Two toxins that are potent inhibitors of protein synthesis, the plant toxin ricin and the bacterial Shiga toxin, are transported from endosomes to the TGN by Rab9-independent mechanisms.[66] Endosomal tubules frequently contain clathrin-coated buds,[67] and clathrin depletion strongly inhibits trafficking of Shiga toxin from endosomes to the TGN, whereas endocytosis of the toxin is less affected.[68,69] In addition to clathrin, the clathrin-binding protein EpsinR is also required for retrograde trafficking of Shiga toxin, whereas the clathrin adaptor AP-1 is not required.[68] Although Rab9 is not required, other Rab GTPases have been found to regulate the retrograde trafficking of Shiga toxin: Rab11, which regulates trafficking through the RE, is involved in the retrograde trafficking of Shiga toxin.[70] While the small GTPase Rab6 has been reported to control retrograde trafficking within the Golgi apparatus, a closely related splice variant, Rab6', has been found to regulate the trafficking of Shiga toxin from endosomes to the TGN.[71] Even though the retrograde trafficking of Shiga toxin is clearly clathrin- and Rab11-dependent, this is not the case with ricin.[72] This indicates that, although both these toxins exploit trafficking pathways from endosomes to the TGN, they follow different trafficking routes. Taken together with the data on M6PR trafficking, this indicates that there are multiple trafficking routes from endosomes to the TGN. This points to the existence of several distinct sorting machineries, of which only few subcomponents are known at the moment.

Conclusions and Perspectives

Accumulating evidence suggests that endosomal cargo molecules may be sorted along several alternative pathways (Fig. 1), specified by signals embedded within the cargo structure. These signals are surprisingly diverse, as are the machineries that recognize them (Table 1). The emerging picture is that there are several parallel pathways that mediate endocytic sorting to the recycling, degradative and biosynthtic pathways, and that multiple specific carriers are involved. Some molecules, such as the M6PRs, are sorted along at least two independent pathways out of endosomes, perhaps in order to ensure efficient retrieval. It is going to be a great challenge to untangle the molecular mechanisms that distinguish various sorting pathways.

For simplicity, we have in this review discussed endosomal sorting in nonpolarized cells such as fibroblasts. The situation is even more complicated in polarized cells such as neurons and epithelial cells. Here, specialized endosomal pathways exist, including the sorting of endocytosed proteins into regenerating synaptic vesicles[73] and the communication between apical and basolateral endosomes.[74] We also have not discussed certain less-characterized pathways, like the direct sorting of endocytosed SV40 virus from endosomes/caveosomes to the endoplasmic reticulum,[75] and the trafficking of MHC class II molecules from LEs/lysosomes to the plasma membrane of dendritic cells.[76] This multitude of endosomal trafficking pathways is likely to be mirrored by an increasing complexity of sorting signals and sorting machineries. These machineries may not only include protein complexes, as discussed here, but also membrane lipids that cluster into "rafts".[77]

One of the best-characterized sorting mechanisms in the endosome involves the degradative sorting of mono- or multi-ubiquitinated membrane proteins by a conserved machinery that involves Hrs and ESCRT proteins.[3,42] However, even in this case, we only have an incomplete picture of the sorting mechanisms. Why are some cargoes ubiquitinated and other not? Even though ubiquitin ligases that append ubiquitin to specific endosomal cargoes have been identified,[78-80] it has been surprisingly difficult to identify consensus sequence motifs that are targeted for ubiquitination. Presumably, the signals that specify ubiquitination are encoded by three-dimensional determinants and therefore difficult to identify by sequence analyses alone. Another conceptual problem that concerns the ubiquitin-recognition machinery is the idea that ubiquitinated cargo is delivered from one protein complex to another. How is this process driven in the right direction? It is possible that interactions between the sorting machinery and ubiquitin are regulated by their phosphorylation or ubiquitination,[81] although this remains to be determined. For other sorting steps, such as retrograde trafficking to the biosynthetic pathway, our knowledge is even less complete at the moment. Nevertheless, even if our view of the sorting signals and machineries is still very fragmentary, the ongoing functional genomic analyses and structural determinations of cargo-machinery interactions are likely to yield substantial new information within a short space of time.

References

1. Zerial M, McBride H. Rab proteins as membrane organizers. Nat Rev Mol Cell Biol 2001; 2:107-117.
2. Maxfield FR, McGraw TE. Endocytic recycling. Nat Rev Mol Cell Biol 2004; 5:121-132.
3. Raiborg C, Rusten TE, Stenmark H. Protein sorting into multivesicular endosomes. Curr Opin Cell Biol 2003; 15:446-455.

4. Gruenberg J, Stenmark H. The biogenesis of multivesicular endosomes. Nat Rev Mol Cell Biol 2004; 5:317-323.
5. Sheff DR, Daro EA, Hull M et al. The receptor recycling pathway contains two distinct populations of early endosomes with different sorting functions. J Cell Biol 1999; 145:123-139.
6. Johnson LS, Dunn KW, Pytowski B et al. Endosome acidification and receptor trafficking: Bafilomycin A1 slows receptor externalization by a mechanism involving the receptor's internalization motif. Mol Biol Cell 1993; 4:1251-1266.
7. Gruenberg J, Maxfield FR. Membrane transport in the endocytic pathway. Curr Opin Cell Biol 1995; 7:552-563.
8. Dai J, Li J, Bos E et al. ACAP1 promotes endocytic recycling by recognizing recycling sorting signals. Dev Cell 2004; 7:771-776.
9. Yan Q, Sun W, Kujala P et al. CART: An Hrs/actinin-4/BERP/myosin V protein complex required for efficient receptor recycling. Mol Biol Cell 2005; 16:2470-2482.
10. Johnson AO, Lampson MA, McGraw TE. A di-leucine sequence and a cluster of acidic amino acids are required for dynamic retention in the endosomal recycling compartment of fibroblasts. Mol Biol Cell 2001; 12:367-381.
11. Sorkin A, Von Zastrow M. Signal transduction and endocytosis: Close encounters of many kinds. Nat Rev Mol Cell Biol 2002; 3:600-614.
12. Cao TT, Deacon HW, Reczek D et al. A kinase-regulated PDZ-domain interaction controls endocytic sorting of the beta2-adrenergic receptor. Nature 1999; 401:286-290.
13. Hanyaloglu AC, McCullagh E, von ZM. Essential role of Hrs in a recycling mechanism mediating functional resensitization of cell signaling. EMBO J 2005; 24:2265-2283.
14. Tanowitz M, Von Zastrow M. A novel endocytic recycling signal that distinguishes the membrane trafficking of naturally occurring opioid receptors. J Biol Chem 2003; 278:45978-45986.
15. Galet C, Hirakawa T, Ascoli M. The postendocytotic trafficking of the human lutropin receptor is mediated by a transferable motif consisting of the C-terminal cysteine and an upstream leucine. Mol Endocrinol 2004; 18:434-446.
16. Vargas GA, Von Zastrow M. Identification of a novel endocytic recycling signal in the D1 dopamine receptor. J Biol Chem 2004; 279:37461-37469.
17. Haglund K, Sigismund S, Polo S et al. Multiple monoubiquitination of RTKs is sufficient for their endocytosis and degradation. Nat Cell Biol 2003; 5:461-466.
18. Mosesson Y, Shtiegman K, Katz M et al. Endocytsosis of receptor tyrosine kinases is driven by mono-, not poly- Ubiquitylation. J Biol Chem 2003.
19. Hicke L, Dunn R. Regulation of membrane protein transport by ubiquitin and ubiquitin-binding proteins. Annu Rev Cell Dev Biol 2003; 19:141-172.
20. Reggiori F, Pelham HR. Sorting of proteins into multivesicular bodies: Ubiquitin-dependent and -independent targeting. EMBO J 2001; 20:5176-5186.
21. Raiborg C, Bache KG, Gillooly DJ et al. Hrs sorts ubiquitinated proteins into clathrin-coated microdomains of early endosomes. Nat Cell Biol 2002; 4:394-398.
22. Dunn R, Hicke L. Domains of the Rsp5 ubiquitin-protein ligase required for receptor-mediated and fluid-phase endocytosis. Mol Biol Cell 2001; 12:421-435.
23. Haglund K, Di Fiore PP, Dikic I. Distinct monoubiquitin signals in receptor endocytosis. Trends Biochem Sci 2003; 28:598-603.
24. Hofmann K, Falquet L. A ubiquitin-interacting motif conserved in components of the proteasomal and lysosomal protein degradation systems. Trends Biochem Sci 2001; 26:347-350.
25. Hofmann K, Bucher P. The UBA domain: A sequence motif present in multiple enzyme classes of the ubiquitination pathway. Trends Biochem Sci 1996; 21:172-173.
26. Katzmann DJ, Babst M, Emr SD. Ubiquitin-dependent sorting into the multivesicular body pathway requires the function of a conserved endosomal protein sorting complex, ESCRT-I. Cell 2001; 106:145-155.
27. Garrus JE, von Schwedler UK, Pornillos OW et al. Tsg101 and the vacuolar protein sorting pathway are essential for HIV-1 budding. Cell 2001; 107:55-65.
28. Shih SC, Prag G, Francis SA et al. A ubiquitin-binding motif required for intramolecular monoubiquitylation, the CUE domain. EMBO J 2003; 22:1273-1281.
29. Donaldson KM, Yin H, Gekakis N et al. Ubiquitin signals protein trafficking via interaction with a novel ubiquitin binding domain in the membrane fusion regulator, Vps9p. Curr Biol 2003; 13:258-262.
30. Hook SS, Orian A, Cowley SM et al. Histone deacetylase 6 binds polyubiquitin through its zinc finger (PAZ domain) and copurifies with deubiquitinating enzymes. Proc Natl Acad Sci USA 2002; 99:13425-13430.

31. Meyer HH, Shorter JG, Seemann J et al. A complex of mammalian ufd1 and npl4 links the AAA-ATPase, p97, to ubiquitin and nuclear transport pathways. EMBO J 2000; 19:2181-2192.
32. Shiba Y, Katoh Y, Shiba T et al. GAT (GGA and Tom1) domain responsible for ubiquitin binding and ubiquitination. J Biol Chem 2004; 279:7105-7111.
33. Slagsvold T, Aasland R, Hirano S et al. Eap45 in mammalian ESCRT-II binds ubiquitin via a phosphoinositide-interacting GLUE domain. J Biol Chem 2005; 280:19600-19606.
34. Raiborg C, Stenmark H. Hrs and endocytic sorting of ubiquitinated membrane proteins. Cell Struct Funct 2002; 27:403-408.
35. Stenmark H, Aasland R, Driscoll PC. The phosphatidylinositol 3-phosphate-binding FYVE finger. FEBS Lett 2002; 513:77-84.
36. Gillooly DJ, Morrow IC, Lindsay M et al. Localization of phosphatidylinositol 3-phosphate in yeast and mammalian cells. EMBO J 2000; 19:4577-4588.
37. Raiborg C, Bremnes B, Mehlum A et al. FYVE and coiled-coil domains determine the specific localisation of Hrs to early endosomes. J Cell Sci 2001; 114:2255-2263.
38. Raiborg C, Bache KG, Mehlum A et al. Hrs recruits clathrin to early endosomes. EMBO J 2001; 20:5008-5021.
39. Sachse M, Urbe S, Oorschot V et al. Bilayered clathrin coats on endosomal vacuoles are involved in protein sorting toward lysosomes. Mol Biol Cell 2002; 13:1313-1328.
40. Bache KG, Raiborg C, Mehlum A et al. STAM and Hrs are subunits of a multivalent Ubiquitin-binding complex on early endosomes. J Biol Chem 2003; 278:12513-12521.
41. Puertollano R, Bonifacino JS. Interactions of GGA3 with the ubiquitin sorting machinery. Nat Cell Biol 2004; 6:244-251.
42. Katzmann DJ, Odorizzi G, Emr SD. Receptor downregulation and multivesicular-body sorting. Nat Rev Mol Cell Biol 2002; 3:893-905.
43. Babst M, Katzmann DJ, Snyder WB et al. Endosome-associated complex, ESCRT-II, recruits transport machinery for protein sorting at the multivesicular body. Dev Cell 2002; 3:283-289.
44. Babst M, Katzmann DJ, Estepa-Sabal EJ et al. Escrt-III: An endosome-associated heterooligomeric protein complex required for mvb sorting. Dev Cell 2002; 3:271-282.
45. Bache KG, Brech A, Mehlum A et al. Hrs regulates multivesicular body formation via ESCRT recruitment to endosomes. J Cell Biol 2003; 162:435-442.
46. Pornillos O, Higginson DS, Stray KM et al. HIV Gag mimics the Tsg101-recruiting activity of the human Hrs protein. J Cell Biol 2003; 162:425-434.
47. Sundquist WI, Schubert HL, Kelly BN et al. Ubiquitin recognition by the human TSG101 protein. Mol Cell 2004; 13:783-789.
48. Teo H, Perisic O, Gonzalez B et al. ESCRT-II, an endosome-associated complex required for protein sorting: Crystal structure and interactions with ESCRT-III and membranes. Dev Cell 2004; 7:559-569.
49. Teo H, Veprintsev DB, Williams RL. Structural insights into endosomal sorting complex required for transport (ESCRT-I) recognition of ubiquitinated proteins. J Biol Chem 2004; 279:28689-28696.
50. Hierro A, Sun J, Rusnak AS et al. Structure of the ESCRT-II endosomal trafficking complex. Nature 2004; 431:221-225.
51. Babst M, Wendland B, Estepa EJ et al. The Vps4p AAA ATPase regulates membrane association of a Vps protein complex required for normal endosome function. EMBO J 1998; 17:2982-2993.
52. Tanowitz M, Von Zastrow M. Ubiquitination-independent trafficking of G protein-coupled receptors to lysosomes. J Biol Chem 2002; 277:50219-50222.
53. Hislop JN, Marley A, Von Zastrow M. Role of mammalian vacuolar protein-sorting proteins in endocytic trafficking of a nonubiquitinated G protein-coupled receptor to lysosomes. J Biol Chem 2004; 279:22522-22531.
54. Whistler JL, Enquist J, Marley A et al. Modulation of postendocytic sorting of G protein-coupled receptors. Science 2002; 297:615-620.
55. Peden AA, Oorschot V, Hesser BA et al. Localization of the AP-3 adaptor complex defines a novel endosomal exit site for lysosomal membrane proteins. J Cell Biol 2004; 164:1065-1076.
56. Ghosh P, Dahms NM, Kornfeld S. Mannose 6-phosphate receptors: New twists in the tale. Nat Rev Mol Cell Biol 2003; 4:202-212.
57. Diaz E, Pfeffer SR. TIP47: A cargo selection device for mannose 6-phosphate receptor trafficking. Cell 1998; 93:433-443.
58. Carroll KS, Hanna J, Simon I et al. Role of Rab9 GTPase in facilitating receptor recruitment by TIP47. Science 2001; 292:1373-1376.
59. Griffiths G. The structure and function of a mannose 6-phosphate receptor-enriched, prelysosomal compartment in animal cells. J Cell Sci Suppl 1989; 11:139-147.

60. Seaman MN, McCaffery JM, Emr SD. A membrane coat complex essential for endosome-to-Golgi retrograde transport in yeast. J Cell Biol 1998; 142:665-681.
61. Arighi CN, Hartnell LM, Aguilar RC et al. Role of the mammalian retromer in sorting of the cation-independent mannose 6-phosphate receptor. J Cell Biol 2004; 165:123-133.
62. Haft CR, de la Luz SM, Bafford R et al. Human orthologs of yeast vacuolar protein sorting proteins Vps26, 29, and 35: Assembly into multimeric complexes. Mol Biol Cell 2000; 11:4105-4116.
63. Carlton J, Bujny M, Peter BJ et al. Sorting nexin-1 mediates tubular endosome-to-TGN transport through coincidence sensing of high- Curvature membranes and 3-phosphoinositides. Curr Biol 2004; 14:1791-1800.
64. Seaman MN. Cargo-selective endosomal sorting for retrieval to the Golgi requires retromer. J Cell Biol 2004; 165:111-122.
65. Seaman MNJ, Marcusson EG, Cereghino JL et al. Endosome to Golgi retrieval of the vacuolar protein sorting receptor, Vps10p, requires the function of the VPS29, VPS30, and VPS35 gene products. J Cell Biol 1997; 137:79-92.
66. Sandvig K, Spilsberg B, Lauvrak SU et al. Pathways followed by protein toxins into cells. Int J Med Microbiol 2004; 293:483-490.
67. Stoorvogel W, Oorschot V, Geuze HJ. A novel class of clathrin-coated vesicles budding from endosomes. J Cell Biol 1996; 132:21-33.
68. Saint-Pol A, Yelamos B, Amessou M et al. Clathrin adaptor epsinR is required for retrograde sorting on early endosomal membranes. Dev Cell 2004; 6:525-538.
69. Lauvrak SU, Torgersen ML, Sandvig K. Efficient endosome-to-Golgi transport of Shiga toxin is dependent on dynamin and clathrin. J Cell Sci 2004; 117:2321-2331.
70. Wilcke M, Johannes L, Galli T et al. Rab11 regulates the compartmentalization of early endosomes required for efficient transport from early endosomes to the trans-golgi network. J Cell Biol 2000; 151:1207-1220.
71. Mallard F, Tang BL, Galli T et al. Early/recycling endosomes-to-TGN transport involves two SNARE complexes and a Rab6 isoform. J Cell Biol 2002; 156:653-664.
72. Iversen TG, Skretting G, Llorente A et al. Endosome to Golgi transport of ricin is independent of clathrin and the Rab9- and Rab11-GTPases. Mol Biol Cell 2001; 12:2099-2107.
73. Murthy VN, De Camilli P. Cell biology of the presynaptic terminal. Annu Rev Neurosci 2003; 26:701-728.
74. Mostov KE, Verges M, Altschuler Y. Membrane traffic in polarized epithelial cells. Curr Opin Cell Biol 2000; 12:483-490.
75. Pelkmans L, Puntener D, Helenius A. Local actin polymerization and dynamin recruitment in SV40-induced internalization of caveolae. Science 2002; 296:535-539.
76. Chow A, Toomre D, Garrett W et al. Dendritic cell maturation triggers retrograde MHC class II transport from lysosomes to the plasma membrane. Nature 2002; 418:988-994.
77. Simons K, Ikonen E. Functional rafts in cell membranes. Nature 1997; 387:569-572.
78. Katzmann DJ, Sarkar S, Chu T et al. Multivesicular body sorting: Ubiquitin ligase Rsp5 is required for the modification and sorting of carboxypeptidase S. Mol Biol Cell 2004; 15:468-480.
79. Marchese A, Raiborg C, Santini F et al. The E3 ubiquitin ligase AIP4 mediates ubiquitination and sorting of the G protein-coupled receptor CXCR4. Dev Cell 2003; 5:709-722.
80. Levkowitz G, Waterman H, Zamir E et al. c-Cbl/Sli-1 regulates endocytic sorting and ubiquitination of the epidermal growth factor receptor. Genes Dev 1998; 12:3663-3674.
81. Polo S, Sigismund S, Faretta M et al. A single motif responsible for ubiquitin recognition and monoubiquitination in endocytic proteins. Nature 2002; 416:451-455.
82. Deneka M, Neeft M, Popa I et al. Rabaptin-5alpha/rabaptin-4 serves as a linker between rab4 and gamma(1)-adaptin in membrane recycling from endosomes. EMBO J 2003; 22:2645-2657.
83. Munshi UM, Clouser CL, Peegel H et al. Evidence that palmitoylation of carboxyl terminus cysteine residues of the human luteinizing hormone receptor regulates post-endocytic processing. Mol Endocrinol 2004.
84. Roquemore EP, Banting G. Efficient trafficking of TGN38 from the endosome to the trans-Golgi network requires a free hydroxyl group at position 331 in the cytosolic domain. Mol Biol Cell 1998; 9:2125-2144.
85. Babst M, Odorizzi G, Estepa EJ et al. Mammalian tumor suceptibility gene 101 (TSG101) and the yeast homologue, Vps23p, both function in late endosomal trafficking. Traffic 2000; 1:248-258.
86. Bishop N, Horman A, Woodman P. Mammalian class E vps proteins recognize ubiquitin and act in the removal of endosomal protein-ubiquitin conjugates. J Cell Biol 2002; 157:91-101.
87. Bache KG, Slagsvold T, Cabezas A et al. The growth-regulatory protein HCRP1/hVps37A is a subunit of mammalian ESCRT-I and mediates receptor down-regulation. Mol Biol Cell 2004; 15:4337-4346.

88. Sharma M, Pampinella F, Nemes C et al. Misfolding diverts CFTR from recycling to degradation: Quality control at early endosomes. J Cell Biol 2004; 164:923-933.

89. Obermuller S, Kiecke C, von Figura K et al. The tyrosine motifs of Lamp 1 and LAP determine their direct and indirect targetting to lysosomes. J Cell Sci 2002; 115:185-194.

90. Schweizer A, Kornfeld S, Rohrer J. Proper sorting of the cation-dependent mannose 6-phosphate receptor in endosomes depends on a pair of aromatic amino acids in its cytoplasmic tail. Proc Natl Acad Sci USA 1997; 94:14471-14476.

91. Krise JP, Sincock PM, Orsel JG et al. Quantitative analysis of TIP47-receptor cytoplasmic domain interactions: Implications for endosome-to-trans Golgi network trafficking. J Biol Chem 2000; 275:25188-25193.

92. Tikkanen R, Obermuller S, Denzer K et al. The dileucine motif within the tail of MPR46 is required for sorting of the receptor in endosomes. Traffic 2000; 1:631-640.

93. Schapiro FB, Soe TT, Mallet WG et al. Role of cytoplasmic domain serines in intracellular trafficking of furin. Mol Biol Cell 2004; 15:2884-2894.

94. Horiuchi H, Lippé R, McBride HM et al. A novel Rab5 GDP/GTP exchange factor complexed to Rabaptin-5 links nucleotide exchange to effector recruitment and function. Cell 1997; 90:1149-1159.

95. Prag G, Misra S, Jones EA et al. Mechanism of ubiquitin recognition by the CUE domain of Vps9p. Cell 2003; 113:609-620.

96. Polo S, Confalonieri S, Salcini AE et al. EH and UIM: Endocytosis and more. Sci STKE 2003; 2003:re17.

97. Shih SC, Katzmann DJ, Schnell JD et al. Epsins and Vps27p/Hrs contain ubiquitin-binding domains that function in receptor endocytosis. Nat Cell Biol 2002; 4:389-393.

98. Prag G, Lee S, Mattera R et al. Structural mechanism for ubiquitinated-cargo recognition by the Golgi-localized, gamma-ear-containing, ADP-ribosylation-factor-binding proteins. Proc Natl Acad Sci USA 2005; 102:2334-2339.

99. Bonifacino JS. The GGA proteins: Adaptors on the move. Nat Rev Mol Cell Biol 2004; 5:23-32.

100. Katoh Y, Shiba Y, Mitsuhashi H et al. Tollip and Tom1 form a complex and recruit ubiquitin-conjugated proteins onto early endosomes. J Biol Chem 2004; 279:24435-24443.

101. Puertollano R. Interactions of TOM1L1 with the multivesicular body sorting machinery. J Biol Chem 2005; 280:9258-9264.

102. Lloyd TE, Atkinson R, Wu MN et al. Hrs regulates endosome invagination and receptor tyrosine kinase signaling in Drosophila. Cell 2002; 108:261-269.

103. Pornillos O, Higginson DS, Stray KM et al. HIV Gag mimics the Tsg101-recruiting activity of the human Hrs protein. J Cell Biol 2003; 162:425-434.

104. Bilodeau PS, Urbanowski JL, Winistorfer SC et al. The Vps27p Hse1p complex binds ubiquitin and mediates endosomal protein sorting. Nat Cell Biol 2002; 4:534-539.

105. Mizuno E, Kawahata K, Kato M et al. STAM proteins bind ubiquitinated proteins on the early endosome via the VHS domain and ubiquitin-interacting motif. Mol Biol Cell 2003; 14:3675-3689.

106. Kanazawa C, Morita E, Yamada M et al. Effects of deficiencies of STAMs and Hrs, mammalian class E Vps proteins, on receptor downregulation. Biochem Biophys Res Commun 2003; 309:848-856.

107. Bilodeau PS, Winistorfer SC, Kearney WR et al. Vps27-Hse1 and ESCRT-I complexes cooperate to increase efficiency of sorting ubiquitinated proteins at the endosome. J Cell Biol 2003; 163:237-243.

108. Alam SL, Sun J, Payne M et al. Ubiquitin interactions of NZF zinc fingers. EMBO J 2004; 23:1411-1421.

109. Bishop N, Woodman P. ATPase-defective mammalian VPS4 localizes to aberrant endosomes and impairs cholesterol trafficking. Mol Biol Cell 2000; 11:227-239.

110. Whitley P, Reaves BJ, Hashimoto M et al. Identification of mammalian Vps24p as an effector of phosphatidylinositol 3,5-bisphosphate-dependent endosome compartmentalization. J Biol Chem 2003; 278:38786-38795.

CHAPTER 8

Signaling from Internalized Receptors

Simona Polo, Letizia Lanzetti and Silvia Giordano*

Abstract

A ctivation of many receptors triggers a cascade of signal transducing events and increases their rate of internalization. Receptor endocytosis has always been viewed primarily as a mechanism to negatively regulate receptor activation, but recent evidence suggests that internalization may result in the formation of specialized signaling platforms on intracellular vesicles. Thus, the investigation of the molecular composition of the various vesicular compartments, their interplay and their spatial and temporal regulation is crucial in order to fully understand the modality of cell signaling.

Introduction

Cells sense and respond to extracellular signals via a dynamic signal-transduction system capable of supporting or inhibiting cell activation. Efficient delivery of signals from the extracellular environment to the intracellular compartment is critical for a tight control of cell growth and differentiation. The first cellular components involved in this signal-transduction system are cell surface receptors, which come in contact with extracellular stimuli, delivered as soluble or membrane bound ligands. It has been known for years that the interaction between receptors and their ligands triggers a cascade of intracellular signals that ultimately leads to cell proliferation. These pathways have been deeply studied and now we have a comprehensive, although not definitive, picture of how they work and of which biological responses are elicited as consequence of their activation. However, only recently it became clear that activated receptors also promote a series of events leading to their endocytosis. Upon internalization, activated receptors are sorted to the endosomes and can be either recycled or degraded in the endosomal or the proteosomal compartments.[1,2] Originally, endocytosis has been considered simply a bowl to eliminate the signaling complexes, but it is now known that internalized receptors are still active and can interact with intracellular transducers and activate new signaling pathways. Since it is currently clear that the output of a transduction process depends not only on the "quality" of the activated signal, but also on its strength and on the location of the emitted signal, new interest on the study of the endocytic trafficking has arisen. The endocytic pathway can achieve signaling compartimentalization since it is organized in a net of distinct but interconnected membrane domains and recent works suggest that it can play a direct role in signal propagation and control.

The rising questions are thus: are the signals originated in the endosomal compartment required for receptors-induced biological responses? Are these signals different from those originated in the plasma membrane? Is there any difference, in term of signal transduction, if the

*Corresponding Author: Silvia Giordano—Institute for Cancer Research and Treatment (IRCC), Division of Molecular Oncology, University of Turin Medical School, Str. Prov. 142, Km 3.95, 10060 Candiolo, Turin, Italy, Email: silvia.giordano@ircc.it

Endosomes, edited by Ivan Dikic. ©2006 Landes Bioscience and Springer Science+Business Media.

receptors are internalized through different pathways (i.e., coated vesicles vs caveolae)? The answers to all these questions are not yet clear, but we are starting to understand that receptor internalization is not simply a way to remove activated receptors from the plasma membrane and that the internalization pathway followed by the receptor can influence the signaling ability of the receptor itself, ultimately leading to different biological responses.[3-6] Moreover, these processes differ from receptor to receptor and the common machinery can be utilized by the different receptors to obtain different outcomes.

In this chapter we will thus describe separately some of the receptors for which the role of internalization on signaling ability is better known.

RTKS: Temporal and Spatial Regulation of Signaling

Receptors for most growth factors are transmembrane proteins endowed with tyrosine kinase activity (receptor tyrosine kinases, RTKs). On ligand binding, RTKs undergo dimerization that results in promotion of their enzymatic activity. RTK-mediated phosphorylation of tyrosine residues on the receptor itself or on intracellular proteins creates binding sites for other proteins containing phosphotyrosine-binding motives.[7] These intracellular substrates can be either enzymes or adaptors or transcriptional factors. The change of enzymatic activity and the modification of membrane lipids occurring as a result of receptor activation originate a network of protein-protein and protein-lipid interactions, phosphorylations and dephosphorylations, modification of cellular compartmentalization, which ultimately leads to changes in gene transcription and to activation of cell proliferation and/or differentiation. As previously mentioned, upon ligand binding and activation, RTKs undergo endocytosis and move through a series of endocytic compartments.

The possibility that endocytic membrane transport has a role in cell signaling has been extensively studied and several experimental evidences now point to this conclusion (reviewed in ref. 4,5,8). One obvious role for endocytosis in signaling could be to provide a spatial regulation of the signaling. One the best example of this function derived from the study of the TrkA system, the receptor for nerve growth factor (NGF), a neurotrophin that functions as a neuronal survival and differentiation factor (Fig. 1A). When NGF is applied selectively to the tip of the axon, the signal has to travel to reach the cell body in order to modify gene transcription; this cannot be achieved by simple diffusion of the signal, but requires microtubule-mediated retrograde transport to the neural cell body of both TrkA and the signaling molecules.[9] Accumulation in the cell bodies of retrogradely transported TrkA and NGF is thus required for neuronal survival while NGF stimulation of the cell body is not sufficient to induce this biological response.[10] Moreover, while NGF and NT-3 (Neurotrophin 3) exert their effect through the same receptor – TrkA - they control unique aspects of neuronal development through differential TrkA internalization and retrograde signaling. NT-3 signals via cell surface TrkA to support axon growth but not survival, whereas NGF, produced by the neuronal final target after synaptic development, supports not only local axon growth, but also survival, anabolic responses and gene expression through retrograde signaling. It has been shown that NGF treatment leads to endocytosis and retrograde accumulation of activated TrkA, Erk1/2 and Akt complexes, whereas NT-3 fails to induce formation of signaling endosomes, but stimulates vesicle trafficking locally within the axon terminal.[11] This differential control of TrkA trafficking ensures that target-derived NGF and not intermediate target-derived NT-3 is solely responsible for retrograde survival signaling.

Always dealing with "spatial regulation of signaling", it has to be considered that RTK signaling is accomplished by several layers of protein-protein interactions. When the RTK becomes activated and undergoes internalization, the full signaling complex is removed from the cell membrane. The result of this event is that only some of the receptor-bound molecules can find their substrates also in the endosomal compartment and thus can continue to signal. For example, the accumulation of the Grb2-SOS complex in endosomes might serve to sustain for a prolonged time the activity of Ras, which can be either constitutively associated with

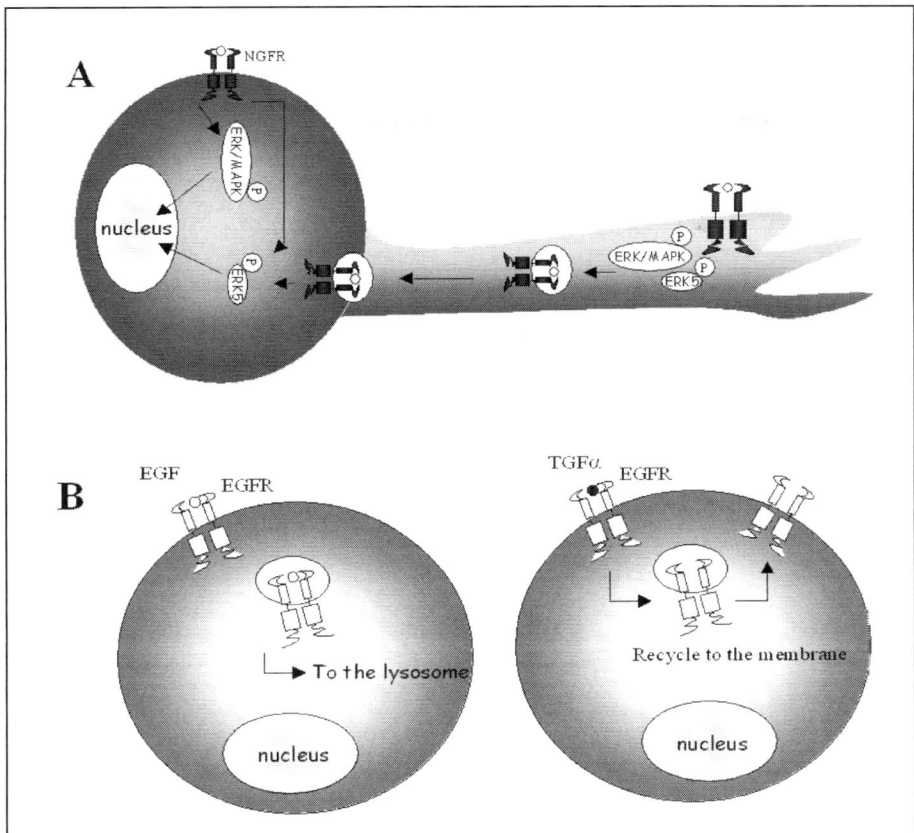

Figure 1. Spatial and temporal regulation of RT signaling. The TrkA system represents one of the best examples of spatial regulation of the signal. When NGF is applied selectively to the tip of the axon, the signal has to travel to reach the cell body in order to modify gene transcription: this cannot be achieved by simple diffusion of the signal but requires microtubule-mediated retrograde transport to the neural cell body of both TrkA and the signaling molecules (modified from: Signal transduction and endocytosis: close encounters of many kinds, Alexander Sorkin and Mark von Zastrow; Nature Reviews Molecular Cell Biology 3, 600-614) B. Endosomes regulate the duration of the signal, controlling the interaction between ligand-receptor pairs. EGFR can interact with several ligands, endowed with different biological abilities. The interaction between EGFR and EGF is quite stable and keeps EGFR in an active state that leads to lysosomal degradation. The instability of the interaction between TGFα and EGFR in the mildly acidic early endosomes does not support receptor sorting into internal multivesicular bodies compartment but rather promotes recycling which results in an increased amount of receptor available on the cell membrane.

endosomes or internalized in response to stimulation. Several effectors of Ras, such as Raf1 and Rab5, are indeed present in the endosomal compartment and can thus be affected by Ras activation.[12-14] On the contrary, when RTK-activated enzymes such as Phospholipase Cγ1 (PLCγ1) and phosphatidyl-inositol 3 kinase (PI3K) are located on the endosomes they are spatially separated from their substrates, which are mainly located at the plasma membrane, and are thus inactive.[15,16] In this way, moving of the RTK signaling complexes to the endosomal compartment can preferentially allow the maintenance of some pathway while inactivating other ones. Another way to differentially assemble protein complexes involves the action of compartment-specific adaptor proteins, capable of linking activated receptors with distinct sets of accessory and effector proteins. As an example, a late endosomal protein, p14, has been shown to be indispensable for epidermal growth factor receptor (EGFR)-mediated efficient ERK activation. In fact, p14 anchors MP1, a MAPK-scaffold protein, to the late endosome,[17] where MP1 specifically binds to MEK1 and thus facilitates activation of Erks. Interference with p14 expression causes displacement of MP1 to the cytosol and prevents full activation of the MEK-Erk cascade, without impairing early Erk1/2 activation occurring at the plasma membrane.

A definitive proof that endosomal signaling of epithelial growth factor receptor stimulates signal transduction pathways leading to biological responses came from the study of Wang and colleagues.[18] By using an experimental system able to dissect signaling originating from the plasma membrane or from the endosomes they showed that the signal transduced from internalized EGFR, with or without a contribution from the plasma membrane, fully satisfy the physiological requirements for S-phase entry.

Another interesting function of endocytosis in signal tranduction is to temporally regulate the length of the signal since its duration is an important parameter to determine the biological outcome. The balance between the number of receptors that undergo degradation versus those that are recycled to the cell membrane is critical to determine the strength of the signal. To this matter, important differences have been observed also among receptors belonging to the same family. The best-studied example is, by far, that of the EGFR family, formed by four structurally related receptors[19] (Fig. 2B). The most oncogenic member of this family, HER2, is poorly downregulated upon activation and recycles very efficiently. On the contrary, EGFR homodimers are effectively directed to a degradative fate following ligand binding, but when EGFR is heterodimerized with HER2 it preferentially undertakes a recycling fate. This event changes the duration of the signaling since new receptors are always exposed on the membrane where they are available for the ligand. Moreover, EGFR signals from endosomes for most of its lifetime, while the other members of the family remain active for longer periods on the plasma membrane.

Endosomes regulate the duration of the signal also controlling the interaction between the ligand-receptor pairs. For example, EGFR can interact with several ligands, endowed with different biological abilities. The stronger mitogenic activity of TGFα (Transforming Growth Factor α) versus EGF is explained on the basis of their differential sensitivity to acidic endosomal pH, that affects the stability of the interaction with EGFR and, thus, its intracellular trafficking.[20] In fact, the instability of the interaction between TGFα and EGFR in the mildly acidic early endosomes does not support receptor sorting into internal multivesicular bodies compartment, but rather promotes recycling which results in an increased amount of receptor available on the cell membrane.

Notch: Endocytosis is Required for Signaling

The family of the Notch receptors represents the prototype of cell surface receptors that require endocytosis in order to signal. Notch signaling plays a fundamental role in regulating cell-fate specification in a variety of developmental and homeostatic processes.[21,22] Alterations in Notch signaling lead to unbalance of these processes and have been implicated in tumorigenesis.[23]

Notch receptors (Notch1-Notch4 in vertebrates) are single-pass membrane receptors that are activated by the Delta/Serrate/Lag2 families of transmembrane ligands, located on the surface of a neighbor cell. The engagement of the extracellular portion of Notch in binding to Delta induces two subsequent proteolytic cleavages[21] (Fig. 2). The first cleavage occurs in the extracellular domain of Notch and it depends on the activity of the TNFα-converting enzyme (TACE).[24,25] The result is the transendocytosis of the Notch extracellular domain/Delta

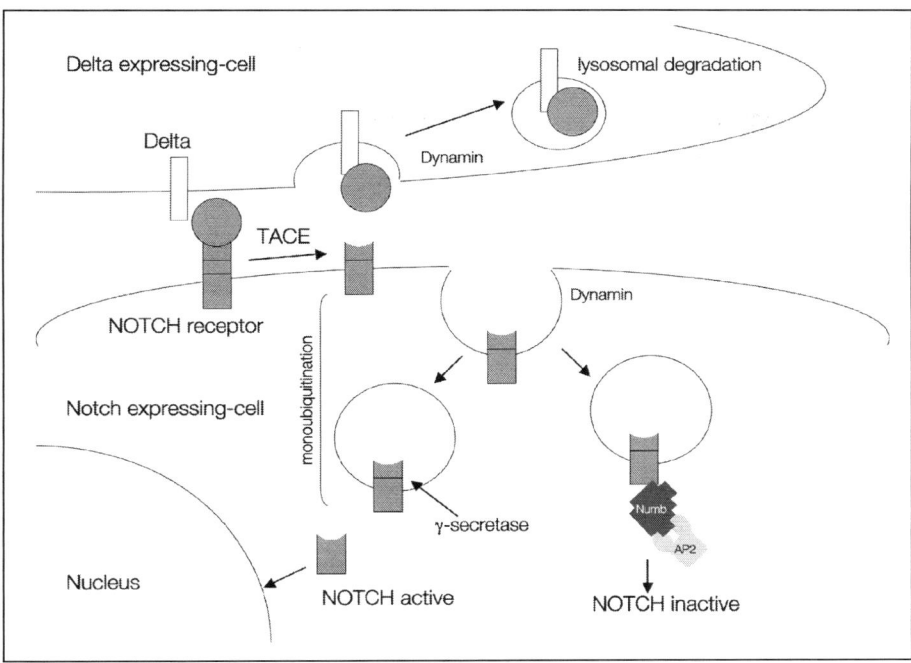

Figure 2. Endocytosis regulates Notch receptor signaling. Notch binds through its extracelluIar domain to Delta that is expressed on the surface of the signal-sending cell. The binding results in the first cleavage operated by the TACE enzyme (TNFα-converting enzyme) in the extrace lular domain of Notch. The Notch extracellular domain is internalized, together with Delta, in the Delta-expressing cell, in a dynamin-dependent manner. The remaining membrane-tethered Notch is cleaved by γ-secretase depending on both monoubiquitination and endocytosis. It causes the release of the intracytoplasmic domain of Notch that translocates into the nucleus where it regulates transcription. Alternatively endocytosis may downregulate Notch activity by removing the receptor from the cell surface via the interaction with the endocytic protein Numb. (modified from Polo S, Pece S, Di Fiore PP. Endocytosis and cancer. Curr Opin Cell Biol. 2004, 2:156-61).

complex into the Delta-expressing cell. This initial endocytic event is a prerequisite for Notch activation in the Notch-expressing cell, since Delta mutants that cannot be internalized, are unable to activate Notch "in vivo".[26] The second cleavage occurs in the transmembrane domain of the remaining membrane-tethered Notch and it is operated by the presenilin/γ-secretase complex. It causes the release of the intracytoplasmatic domain of Notch that translocates into the nucleus where it regulates transcription[27,28] (Fig. 2). Indeed it has been demonstrated that inhibition of endocytosis, by means of dominant-negative mutants such as Dynamin II K44A or Eps15DN, prevents the translocation of the Notch intracytoplasmatic domain into the nucleus.[29,30] Therefore endocytosis is required for proper Notch signaling both in the ligand-expressing cell as well as in the signal-receiving cell.[29]

In addition, Notch is post-translationally modified upon appendage of an ubiquitin moiety to a lysine residue in the juxtamembrane region.[30] Presenilins are able to interact with the monoubiquitinated form of Notch and this step is required for the γ-secretase cleavage.[30] As a pool of presenilins has been found at the cell surface and in the endocytic compartments[31 32] it is conceivable that monoubiquitination and endocytosis of the receptor are necessary for driving Notch to compartments where γ-secretase cleavage can operate.[30]

Endocytosis is therefore acting as a "positive" mechanism in promoting the delivery of the Notch signal to the cell. Nevertheless, Notch signaling is also controlled by endocytosis through a more "classical" mechanism, namely receptor downregulation. A critical player in influencing the Notch availability at the plasma membrane is represented by Numb, an endocytic protein that binds to α-adaptin and localizes to the endosomes in mammalian cells.[33] Numb physically interacts and antagonizes Notch[33,34] as demonstrated in the sensor organ precursor cell of *Drosophila,* where asymmetric partition of Numb and α adaptin at mitosis results in Notch silencing in the Numb-receiving cell and, consequently, in the acquisition of different cell fate.[35] The Numb/Notch antagonism is relevant also in mammals, in particular in the control of tumor proliferation. Increased Notch signaling is observed in Numb-negative tumors, where it can be reverted to basal levels after enforced expression of Numb.[36]

In conclusion, endocytosis regulates Notch signaling at multiple levels: i) modulating Notch signaling via the transendocytosis of the Notch extracellular domain/Delta complex into the ligand presenting cell; ii) controlling the γ-secretase cleavage and therefore the release of the Notch intracellular domain; iii) down-regulating Notch signaling by affecting receptor availability at the plasma membrane.

Different Endocytic Route Different Signaling: The TGF-βR Paradigm

Until recently, it was widely accepted that all cell-surface receptors follow the same endocytic pathway and that they are internalized by clathrin-coated pits, with sorting at the cell surface being achieved solely through the direct or indirect binding of receptor cytoplasmic domains to clathrin associated proteins. However, clathrin-independent ways of entry into the cell also exist. In particular, progress has been made in characterizing a pathway involving cholesterol- and glycosphingolipid-rich membrane domains (rafts) and in identifying their cargoes.[37] It was shown that the TGF-β (Transforming Growth Factor β) internalizes through both coated pits and caveolae. The former route is associated with increased receptor signaling from early endosomes, while the caveolar pathway causes rapid receptor degradation.[38] Thus, depending on the entry route, the fate of internalized TGF-β receptors will be different (Fig. 3).

At the molecular level activated TGF-β receptors bind and phosphorylate R-Smads (Smad2 and Smad3). R-Smad binding to the receptors is facilitated by a protein called SARA (Smad Anchor for Receptor Activation) which has a FYVE domain and is predominantly localized to the PitdIns3P-enriched endosomes, a key compartment for Smad activation.[39] Clathrin-dependent internalization of the TGF-β receptor into EEA1-positive early endosomes enables the engagement of TGF-β receptor by SARA, necessary to phosphorylate Smad2 and to achieve subsequent propagation of the signal (Fig. 3A). Indeed, interfering with clathrin-dependent trafficking, using K$^+$ depletion, or dominant negative mutants of dynamin or eps15 blocks TGF-β-induced Smad2 activation and nuclear translocation.[38]

On the other end, the other pool of the receptor is internalized via clathrin-independent route, namely a lipid raft–caveolar internalization pathway, which involves caveolin-1-positive vesicles called caveosome. In caveosomes, TGF-β receptors do not encounter SARA, but the ubiquitin ligase Smad7–Smurf2 complex that colocalizes with caveolin-1 and preferentially associates with receptors in rafts.[38] Indeed, TGF-β receptor turnover is inhibited by lipid rafts disruption and this strongly indicates that this pathway leads to degradation and counteracts the clathrin–mediated signaling (Fig. 3B).

Therefore, it seems that Smad signaling components are segregated into two internalization pathways, where they differentially regulate TGF-β signaling. Early endosomes represent a signaling center that functions both to sequester active receptor complexes from rafts and caveolin and to promote access to the Smad2 substrate via SARA, whereas the raft–caveolin compartment

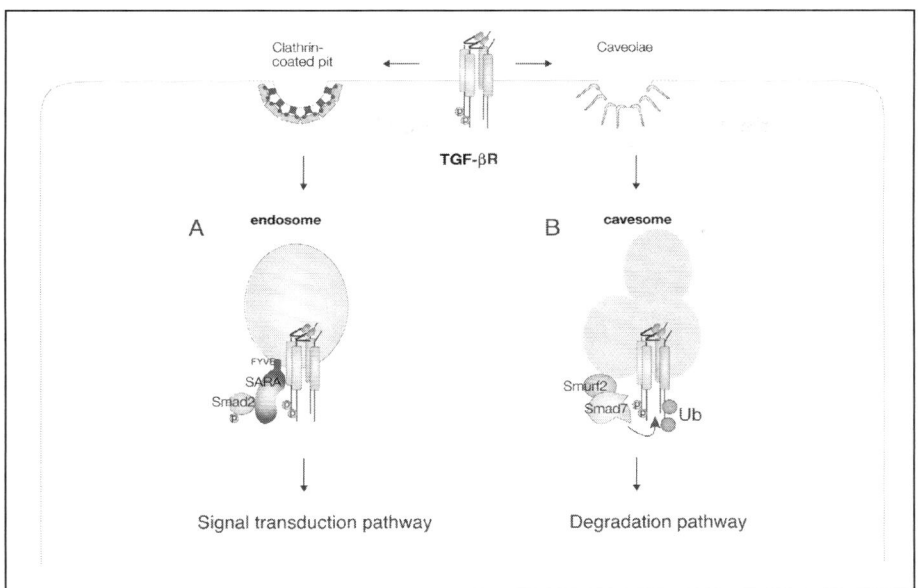

Figure 3. Alternative entry routes for TGF-β receptor. A In the clathrin-mediated pathway, receptor is directed towards the early endosomes, where it interact with Smad2 and SARA. From these vesicles, TGF-βR is able to signal and is recycled back to the cell surface. B In the caveolar-mediated pathway, receptor is directed towards the caveosomes, where it encounter the ubiquitin l gase Smad7–Smurf2 complex and become ubiquitinated and degraded.

may represent a previously undescribed compartment that regulates ubiquitin-dependent degradation of membrane receptors.

An intriguing observation is that expression of constitutively inactive Rab5 (Rab5S34N) stimulates TGF-β signaling, while expression of constitutively active Rab5 (Rab5Q79L) has no effect.[40] One possibility is that Rab5 inhibition may modify the transport of TGF-β receptors between caveosomes and early endosomes, thus underscoring the influence of signaling on trafficking.

Recently, it was also shown that the EGF receptor internalizes through both coated pits and caveolae depending on the level of activation (Sigismund et al, PNAS in press). When the receptor is stimulated with low doses of EGF, is internalized almost exclusively through the clathrin pathway, and it is not ubiquitinated. At higher concentrations of ligand, however, a substantial fraction of the receptor is endocytosed through a lipid raft/caveolar-dependent route, as the receptor becomes ubiquitinated. Interestingly, at low levels of EGF, the EGFR is already fully competent of signaling via its effectors, whereas at high levels of EGF (when clathrin-independent endocytosis becomes significant) there is no increase in signaling abilities, but readily detectable increase in EGFR downregulation. Thus, the combined analysis of TGF-βR and EGFR data suggest the intriguing possibility that the caveolar/raft internalization does not contribute to signaling, and it is preferentially associated with receptor degradation, via ubiquitination.

GPCR and β-Arrestin: Signaling from the Endosomes

Signaling mediated by GPCRs (G protein-coupled receptors) begins with the activation of the receptor through the binding of agonist; this leads to a conformational change within the receptor intracellular domains, which can then be recognized by intracellular proteins.[41] The most common GPCR signal transducing proteins are the heterotrimeric G proteins, which in turn can activate a wide spectrum of effector molecules, including phosphodiesterases, phospholipases, adenylyl cyclases and ion channels.[42] Active GPCRs are also the target of G protein-coupled receptor kinases, which phosphorylate the receptors leading to the rapid recruitment and binding of cytosolic arrestins (known as β-arrestin-1 and -2).[43]

β-arrestins are key players for receptor desensitization and internalization since β-arrestin binding to GPCRs both uncouples receptors from heterotrimeric G proteins and targets them to clathrin-coated pits for endocytosis[44-46] (Fig. 4A). In addition to this established role, the isolation from intact cells of protein complexes containing specific GPCRs, β-arrestins and either Src or ERK/MAPKs has led to the hypothesis that endosome-associated β-arrestins function as molecular scaffolds for the assembly of specific kinase cascades and, perhaps, for the recruitment of other signaling molecules[47,48] that mediate GPCR signaling from endosomes (Fig. 4B). This complex, in fact, has been proposed to mediate a distinct "second wave" of signal transduction through MAPKs, which occurs after the "classical" pathways (such as signaling via adenylyl cyclase) have already been originated at the cell surface.[49] Proteinase-activated receptor 2 (PAR2),[48] angiotensin AT1A receptor (AT1AR)[47] and neurokinin receptor 1 (NK1R)[50] are examples of GPCR for which scaffolding interactions that involve β-arrestins in the ensosomes have been described (Fig. 4B).

Interestingly, activation of MAPK by the heptahelical μ opioid receptor does not require endocytosis of the heptahelical receptor,[51] yet MAPK activation by this receptor is strongly inhibited by a dominant-negative mutant dynamin that blocks endocytosis via coated pits.[51,52] This indicates that dynamin-dependent endocytosis of another molecule might be required for ERK/MAPK signaling by GPCRs, or that dynamin might mediate another function in signaling that is independent of endocytosis per se. Indeed, in some cases, ERK/MAPK signaling by GPCRs is mediated by transactivation of an RTK such as the EGF receptor,[53] and endocytosis of the RTK — but not the GPCR — could be the crucial event that is involved in signal transduction.

It has also been observed that overexpression of mutant dynamin does not detectably affect receptor-mediated activation of Ras and Raf, whereas phosphorylation of ERK/MAPK 1/2 by MEK is strongly inhibited.[54] Therefore, it was postulated that the endosomal localization of MEK, perhaps by endocytic transport of a putative MEK membrane anchor, is essential for ERK/MAPK 1/2 activation. These considerations have led to a search for more specific ways of manipulating endocytic transport of a specific receptor or signaling molecule. One approach has been to examine the effects of mutant βarrestins, which specifically inhibit GPCRs endocytosis without affecting endocytosis of other molecules such as RTKs.[55] However βarrestins, like mutant dynamin, can engage numerous interactions, including those with scaffold signaling kinases,[47,56] so this originates some ambiguities in functional studies of GPCR signaling performed with the use of the inhibitors currently available.

Conclusions

A growing body of evidence points out the role of key signaling molecules in modulating multiple and different signal transduction pathways. The capability to transduce distinct signals relies mainly on the spatial and/or temporal regulation of the molecules. The plasma membrane has long been considered the major signaling-emanating compartment due to the presence of the transmembrane receptors that are engaged by extracellular ligands and in turn initiate the signaling cascade by recruiting downstream mediators and by activating GTPases and other enzymes. Nevertheless, it has been demonstrated that signaling platforms are also assembled on early endosomes and that they can propagate signals inside the cell. In addition other intracellular membrane compartments appear to be responsible for signal propagation like for instance the Golgi apparatus. The Golgi is involved in protein maturation and secretion, thus controlling the inside-out signaling. The discovery that the GTPase Ras (Ha-Ras) localizes at the Golgi and signals from this compartment reveals a role for the Golgi also in the

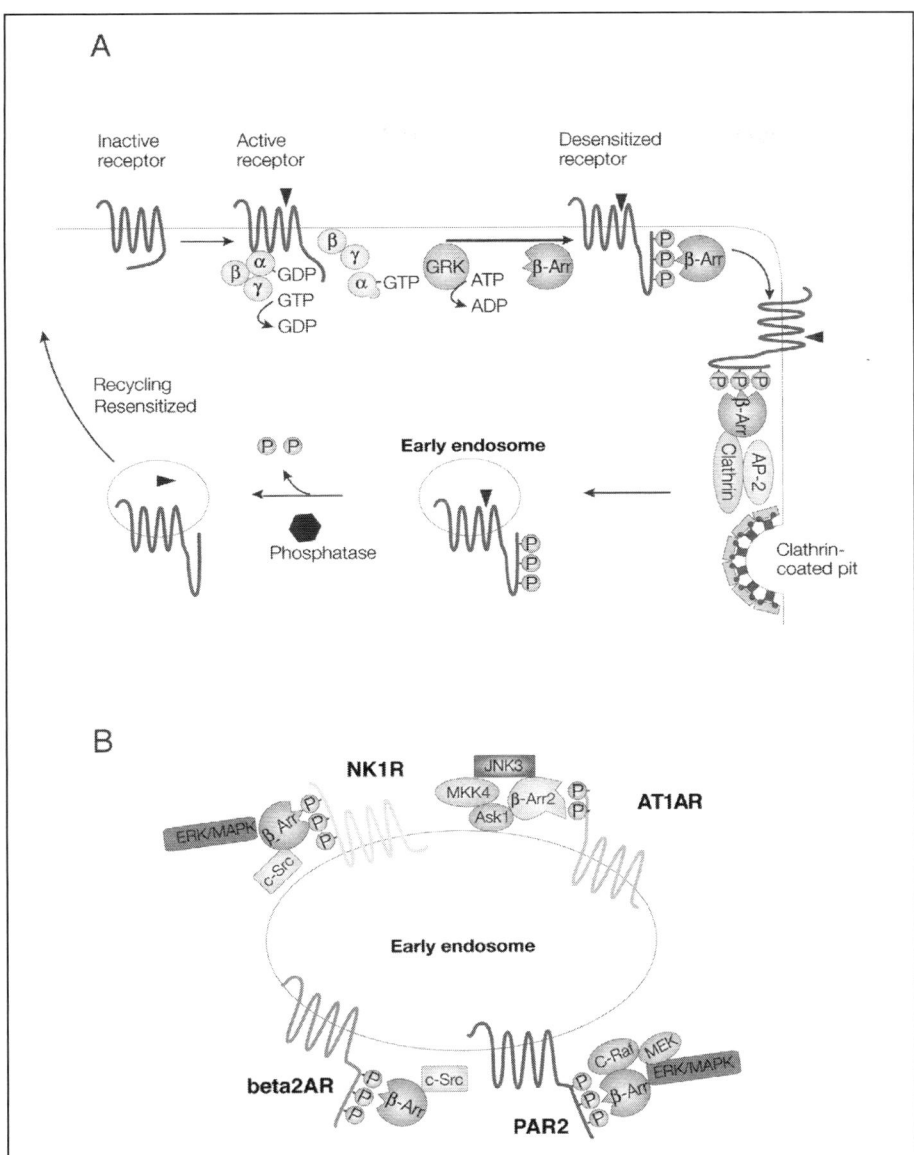

Figure 4. GPCR activity at the membrane and in the endosomes. A) At the membrane activated GPCR interact with the heterotrimeric G proteins, which in turn, activate the signaling cascade. Upon activation GPCR are desensitized by phosphorylation and interact with β-arrestins. β-arrestins promote receptor internalization via clathrin-mediated pathway into endosomes where receptors are dephosphorylated and recycled back to the membrane (resensitization). B) In the endosomes GPCR, through β-arrestins, can interact with various component of the signaling cascade, such as Src and ERK/MAPK kinases, leading to a "second wave" of signal transduction. (modified from: Signal transduction and endocytosis: close encounters of many kinds, Alexander Sorkin and Mark von Zastrow; Nature Reviews Molecular Cell Biology 3, 600-614).

outside-in signaling.[57,58] Therefore the localization or the relocalization of signaling molecules to vesicular compartments is a way to increase signal complexity.

Although the spatial regulation of signaling molecules cannot be, probably, distinguished from a "temporal" regulation, compartmentalization remains an important mechanism of controlling signaling. The temporal regulation of vesicles-derived signals represents an active field of investigation in particular concerning the requirement for vesicular trafficking during the process of cell division (reviewed in ref. 59).

In conclusion the new findings discussed in this chapter argue for a complex, bidirectional cross-talk between signaling and membrane-transport network. Further investigation of the molecular composition of the various vesicular compartments, their interplay and their spatial and temporal regulation is needed in order to clarify the modality of cell signaling.

References

1. Hunter T. Signaling—2000 and beyond. Cell 2000; 100(1):113-127.
2. Waterman H, Yarden Y. Molecular mechanisms underlying endocytosis and sorting of ErbB receptor tyrosine kinases. FEBS Lett 2001; 490(3):142-152.
3. Di Fiore PP, De Camilli P. Endocytosis and signaling: An inseparable partnership. Cell 2001; 106(1):1-4.
4. Sorkin A, Von Zastrow M. Signal transduction and endocytosis: Close encounters of many kinds. Nat Rev Mol Cell Biol 2002; 3(8):600-614.
5. Gonzalez-Gaitan M, Stenmark H. Endocytosis and signaling: A relationship under development. Cell 2003; 115(5):513-521.
6. Miaczynska M, Pelkmans L, Zerial M. Not just a sink: Endosomes in control of signal transduction. Curr Opin Cell Biol 2004; 16(4):400-406.
7. Songyang Z, Shoelson SE, Chaudhuri M et al. SH2 domains recognize specific phosphopeptide sequences. Cell 1993; 72(5):767-778.
8. Gonzalez-Gaitan M. Signal dispersal and transduction through the endocytic pathway. Nat Rev Mol Cell Biol 2003; 4(3):213-224.
9. Howe CL, Valletta JS, Rusnak AS et al. NGF signaling from clathrin-coated vesicles: Evidence that signaling endosomes serve as a platform for the Ras-MAPK pathway. Neuron 2001; 32(5):801-814.
10. Ye H, Kuruvilla R, Zweifel LS et al. Evidence in support of signaling endosome-based retrograde survival of sympathetic neurons. Neuron 2003; 39(1):57-68.
11. Howe CL, Mobley WC. Signaling endosome hypothesis: A cellular mechanism for long distance communication. J Neurobiol 2004; 58(2):207-216.
12. Rizzo MA, Shome K, Watkins SC et al. The recruitment of Raf-1 to membranes is mediated by direct interaction with phosphatidic acid and is independent of association with Ras. J Biol Chem 2000; 275(31):23911-23918.
13. Pol A, Calvo M, Enrich C. Isolated endosomes from quiescent rat liver contain the signal transduction machinery. Differential distribution of activated Raf-1 and Mek in the endocytic compartment. FEBS Lett 1998; 441(1):34-38.
14. Barbieri MA, Roberts RL, Gumusboga A et al. Epidermal growth factor and membrane trafficking. EGF receptor activation of endocytosis requires Rab5a. J Cell Biol 2000; 151(3):539-550.
15. Haugh JM, Schooler K, Wells A et al. Effect of epidermal growth factor receptor internalization on regulation of the phospholipase C-gamma1 signaling pathway. J Biol Chem 1999; 274(13):8958-8965.
16. Haugh JM, Meyer T. Active EGF receptors have limited access to PtdIns(4,5)P(2) in endosomes: Implications for phospholipase C and PI 3-kinase signaling. J Cell Sci 2002; 115(Pt 2):303-310.
17. Teis D, Wunderlich W, Huber LA. Localization of the MP1-MAPK scaffold complex to endosomes is mediated by p14 and required for signal transduction. Dev Cell 2002; 3(6):803-814.
18. Pennock S, Wang Z. Stimulation of cell proliferation by endosomal epidermal growth factor receptor as revealed through two distinct phases of signaling. Mol Cell Biol 2003; 23(16):5803-5815.
19. Marmor MD, Yarden Y. Role of protein ubiquitylation in regulating endocytosis of receptor tyrosine kinases. Oncogene 2004; 23(11):2057-2070.
20. French AR, Tadaki DK, Niyogi SK et al. Intracellular trafficking of epidermal growth factor family ligands is directly influenced by the pH sensitivity of the receptor/ligand interaction. J Biol Chem 1995; 270(9):4334-4340.
21. Mumm JS, Kopan R. Notch signaling: From the outside in. Dev Biol 2000; 228(2):151-165.
22. Artavanis-Tsakonas S, Rand MD, Lake RJ. Notch signaling: Cell fate control and signal integration in development. Science 1999; 284(5415):770-776.

23. Polo S, Pece S, Di Fiore PP. Endocytosis and cancer. Curr Opin Cell Biol 2004; 16(2):156-161.
24. Brou C, Logeat F, Gupta N et al. A novel proteolytic cleavage involved in Notch signaling: The role of the disintegrin-metalloprotease TACE. Mol Cell 2000; 5(2):207-216.
25. Mumm JS, Schroeter EH, Saxena MT et al. A ligand-induced extracellular cleavage regulates gamma-secretase-like proteolytic activation of Notch1. Mol Cell 2000; 5(2):197-206.
26. Parks AL, Klueg KM, Stout JR et al. Ligand endocytosis drives receptor dissociation and activation in the Notch pathway. Development 2000; 127(7):1373-1385.
27. Struhl G, Adachi A. Nuclear access and action of notch in vivo. Cell 1998; 93(4):649-660.
28. Schroeter EH, Kisslinger JA, Kopan R. Notch-1 signalling requires ligand-induced proteolytic release of intracellular domain. Nature 1998; 393(6683):382-386.
29. Seugnet L, Simpson P, Haenlin M. Requirement for dynamin during Notch signaling in Drosophila neurogenesis. Dev Biol 1997; 192(2):585-598.
30. Gupta-Rossi N, Six E, LeBail O et al. Monoubiquitination and endocytosis direct gamma-secretase cleavage of activated Notch receptor. J Cell Biol 2004; 166(1):73-83.
31. Lah JJ, Levey AI. Endogenous presenilin-1 targets to endocytic rather than biosynthetic compartments. Mol Cell Neurosci 2000; 16(2):111-126.
32. Ray WJ, Yao M, Mumm J et al. Cell surface presenilin-1 participates in the gamma-secretase-like proteolysis of Notch. J Biol Chem 1999; 274(51):36801-36807.
33. Santolini E, Puri C, Salcini AE et al. Numb is an endocytic protein. J Cell Biol 2000; 151(6):1345-1352.
34. Guo M, Jan LY, Jan YN. Control of daughter cell fates during asymmetric division: Interaction of Numb and Notch. Neuron 1996; 17(1):27-41.
35. Berdnik D, Torok T, Gonzalez-Gaitan M et al. The endocytic protein alpha-Adaptin is required for numb-mediated asymmetric cell division in Drosophila. Dev Cell 2002; 3(2):221-231.
36. Pece S, Serresi M, Santolini E et al. Loss of negative regulation by Numb over Notch is relevant to human breast carcinogenesis. J Cell Biol 2004; 167(2):215-221.
37. Conner SD, Schmid SL. Regulated portals of entry into the cell. Nature 2003; 422(6927):37-44.
38. Di Guglielmo GM, Le Roy C, Goodfellow AF et al. Distinct endocytic pathways regulate TGF-beta receptor signalling and turnover. Nat Cell Biol 2003; 5(5):410-421.
39. Hayes S, Chawla A, Corvera S. TGF beta receptor internalization into EEA1-enriched early endosomes: Role in signaling to Smad2. J Cell Biol 2002; 158(7):1239-1249.
40. Panopoulou E, Gillooly DJ, Wrana JL et al. Early endosomal regulation of Smad-dependent signaling in endothelial cells. J Biol Chem 2002; 277(20):18046-18052.
41. Hunyady L, Vauquelin G, Vanderheyden P. Agonist induction and conformational selection during activation of a G-protein-coupled receptor. Trends Pharmacol Sci 2003; 24(2):81-86.
42. Hamm HE. The many faces of G protein signaling. J Biol Chem 1998; 273(2):669-672.
43. Miller WE, Lefkowitz RJ. Expanding roles for beta-arrestins as scaffolds and adapters in GPCR signaling and trafficking. Curr Opin Cell Biol 2001; 13(2):139-145.
44. Prossnitz ER. Novel roles for arrestins in the post-endocytic trafficking of G protein-coupled receptors. Life Sci 2004; 75(8):893-899.
45. Claing A, Laporte SA, Caron MG et al. Endocytosis of G protein-coupled receptors: Roles of G protein-coupled receptor kinases and beta-arrestin proteins. Prog Neurobiol 2002; 66(2):61-79.
46. Laporte SA, Oakley RH, Zhang J et al. The beta2-adrenergic receptor/betaarrestin complex recruits the clathrin adaptor AP-2 during endocytosis. Proc Natl Acad Sci USA 1999; 96(7):3712-3717.
47. McDonald PH, Chow CW, Miller WE et al. Beta-arrestin 2: A receptor-regulated MAPK scaffold for the activation of JNK3. Science 2000; 290(5496):1574-1577.
48. DeFea KA, Zalevsky J, Thoma MS et al. beta-arrestin-dependent endocytosis of proteinase-activated receptor 2 is required for intracellular targeting of activated ERK1/2. J Cell Biol 2000; 148(6):1267-1281.
49. Luttrell LM, Lefkowitz RJ. The role of beta-arrestins in the termination and transduction of G-protein-coupled receptor signals. J Cell Sci 2002; 115(Pt 3):455-465.
50. DeFea KA, Vaughn ZD, O'Bryan EM et al. The proliferative and antiapoptotic effects of substance P are facilitated by formation of a beta -arrestin-dependent scaffolding complex. Proc Natl Acad Sci USA 2000; 97(20):11086-11091.
51. Whistler JL, von Zastrow M. Dissociation of functional roles of dynamin in receptor-mediated endocytosis and mitogenic signal transduction. J Biol Chem 1999; 274(35):24575-24578.
52. Ignatova EG, Belcheva MM, Bohn LM et al. Requirement of receptor internalization for opioid stimulation of mitogen-activated protein kinase: Biochemical and immunofluorescence confocal microscopic evidence. J Neurosci 1999; 19(1):56-63.
53. Daub H, Weiss FU, Wallasch C et al. Role of transactivation of the EGF receptor in signalling by G-protein-coupled receptors. Nature 1996; 379(6565):557-560.

54. Kranenburg O, Verlaan I, Moolenaar WH. Dynamin is required for the activation of mitogen-activated protein (MAP) kinase by MAP kinase kinase. J Biol Chem 1999; 274(50):35301-35304.
55. Luttrell LM, Ferguson SS, Daaka Y et al. Beta-arrestin-dependent formation of beta2 adrenergic receptor-Src protein kinase complexes. Science 1999; 283(5402):655-661.
56. McDonald PH, Lefkowitz RJ. Beta-Arrestins: New roles in regulating heptahelical receptors' functions. Cell Signal 2001; 13(10):683-689.
57. Bivona TG, Perez De Castro I, Ahearn IM et al. Phospholipase Cgamma activates Ras on the Golgi apparatus by means of RasGRP1. Nature 2003; 424(6949):694-698.
58. Bivona TG, Philips MR. Ras pathway signaling on endomembranes. Curr Opin Cell Biol 2003; 15(2):136-142.
59. Strickland LI, Burgess DR. Pathways for membrane trafficking during cytokinesis. Trends Cell Biol 2004; 14(3):115-118.

CHAPTER 9

Endocytosis of Receptor Tyrosine Kinases:
Implications for Signal Transduction by Growth Factors

Gal Gur, Yaara Zwang and Yosef Yarden*

Abstract

G rowth factors and their respective receptor tyrosine kinases (RTKs) play pivotal roles in normal cellular functions, such as proliferation and motility, as well as in pathogenesis, including cancer. The amplitude and kinetics of growth factor signaling are determined mainly by a highly regulated endocytic process, which sorts activated receptors to degradation in lysosomes. Molecular mechanisms underlying receptor down-regulation are being unraveled: the active receptor recruits Cbl ubiquitin ligases that decorate it with multiple monomers of ubiquitin. In parallel, Nedd4/AIP4 ubiquitin ligases attach ubiquitin to a set of ubiquitin-binding adaptors (e.g., Epsin and the EGF-receptor protein substrate, Eps15) necessary for the assembly of a clathrin coat. Analogous but distinct ubiquitin-binding platforms underlie receptor sorting into shuttling vesicles at the plasma membrane, early endosomes and a prelysosomal compartment called the multi-vesicular body. In addition to ubiquitylation, phopshorylation of both RTKs and coat adaptors orchestrate receptor sorting in concert with machineries responsible for membrane bending and vesicle fusion. The default route diverts internalized receptors back to the plasma membrane, thus enabling prolonged signaling associated with pathological processes. This review concentrates on the epidermal growth factor receptor (EGFR) as a prototype and highlights the major events occurring on its journey to the lysosome.

Introduction

Cell fate determination in embryogenesis, as well as morphogenic processes throughout adulthood, are regulated primarily by polypeptide growth factors. The initial event underlying stimulation of a target cell by growth factors involves their binding to transmembrane receptor tyrosine kinases (RTKs), whose intracellular portions share a catalytic kinase activity specific for tyrosine residues (reviewed in refs. 1-3). Upon binding to the respective growth factor molecule, RTKs undergo dimerization and catalytic stimulation. This enables them to recruit and/or phosphorylate multiple protein substrates, many of which utilize intrinsic phosphotyrosine-binding regions [e.g., Src-homolgy 2 domain (SH2) and phosphotyrosine-binding domain (PTB)] to bind with phosphorylated tyrosine residues of the active receptor.[4] These receptor-centered events instigate a large number of simultaneous biochemical cascades, which collectively transmit extracellular signals to target organelles and culminate in gross cellular alterations (e.g., cell division or migration). Concomitant with signal distribution and propagation, a variety of desensitization processes are launched, such that the balance between positively-acting and

*Corresponding Author: Yosef Yarden—Department of Biological Regulation, The Weizmann Institute of Science, Rehovot 76100, Israel. Email: yosef.yarden@weizmann.ac.il

Endosomes, edited by Ivan Dikic. ©2006 Landes Bioscience and Springer Science+Business Media.

negatively-acting cascades determines the amplitude and duration of the ensuing biochemical signals (reviewed in ref. 5). Perturbations of this delicate balance often lead to pathogenesis, such as skeletal disorders, cancer and diabetes. We concentrate below on the epidermal growth factor receptor (EGFR) as a prototypic RTK. After briefly describing positively-acting pathways, this chapter will concentrate on the major negatively-acting regulatory process of RTKs, namely: growth factor-induced internalization of active receptors and their subsequent sorting to intracellular degradation.[6-8]

Positively-Acting Signaling Pathways

The most characterized signaling pathways induced upon activation of RTKs, including members of EGFR family (ErbB/HER), are the Ras-mitogen-activated protein kinase (Ras-MAPK), the phosphatidylinositol 3' kinase-protein kinase B (PI3K-PKB/Akt), and the phospholipase C-protein kinase C (PLC-PKC) pathways. The four ErbB proteins, and in fact most RTKs, couple to activation of the Ras-MAPK pathway through SH2 domain-mediated recruitment of the Grb2 adaptor,[9] or indirectly through PTB domain-mediated binding of the Shc adaptor.[10] Regardless of the exact route, active MAPK/Erk molecules translocate to the nucleus to phosphorylate specific transcription factors, such as Sp 1, E2F, Elk-1 and AP1.[11] In a similar manner, receptor phosphorylation provides acceptor sites for the SH2 domain of the regulatory subunit of PI3K, p85. For example, binding of p85 to tyrosine phosphorylated ErbB proteins (predominantly ErbB-3) results in activation of p110, the catalytic subunit of PI3K.[12] Akt (also called protein kinase B, PKB) is a key effector of PI3K; it is recruited to the plasma membrane through its PH domain and activated upon phosphorylation by serine/threonine kinases. The importance of the proliferation and cell survival signals, which are mediated by PI3K, is reflected by the tumor suppressive effects of PTEN. This lipid phosphatase dephosphorylates phosphatidylinositol 3,4,5-trisphosphate [PI(3,4,5)-P$_3$], a lipid required for PKB/Akt activation, and undergoes frequent mutational inactivation in human cancer, which results in constitutive activation of Akt.[13] The third cascade leads to activation of protein kinase C (PKC). Phospholipase Cγ (PLCγ) is recruited to the membrane through SH2 domain-mediated binding to activated RTKs, including EGFR and ErbB-2, as well as through binding of its PH domain to PI3K products (reviewed by ref. 14). Subsequent to phosphorylation by RTKs, PLCγ hydrolyzes PI(4,5)P$_2$ to generate inositol 1,4,5-trisphosphate and 1,2-diacylglycerol (DAG), which are implicated in the mobilization of intracellular calcium ions and activation of PKC, respectively. Several additional signaling pathways are induced by RTKs, including cytoplasmic tyrosine kinases of the Src family and recruitment of transcription factors belonging to the STAT family. Subsequent to their phopshorylation, STAT proteins translocate to the nucleus to activate gene transcription critical for cell proliferation and angiogenesis.

Negatively-Acting Signaling Pathways

Much of the information relevant to attenuation of RTK signaling emerged from studies of invertebrate systems. A single EGFR orthologue and a single EGF-like ligand are found in the worm *C. elegans*, and their signaling is attenuated by a small group of proteins: a Clathrin adaptor, Sli-1 (a c-Cbl orthologue), a GTPase-activating protein called GAP-1 and a cytoplasmic tyrosine kinase homologous to mammalian Ack-1.[15] The single ErbB orthologue of flies is attenuated by similar mechanisms, along with several additional pathways such as an inhibitory ligand, Argos, a family of transmembrane molecules called Kekkons and an adaptor molecule, Sprouty, which has four orthologues in mammals (reviewed by ref. 16). Mammalian RTK signaling is intercepted at multiple additional junctures, which include inhibition of kinase activity by RALT/Mig-6,[17,18] a soluble receptor variant that intercepts receptor activation,[19] a transmembrane inhibitory protein,[20] and specific protein tyrosine phosphatases. Most remarkable is a large family of dual specificity phosphatases called MKPs, which dephosphorylate specific MAP-kinases.[21] It is notable that expression of both MKPs and RALT/Mig-6, along with additional negative regulators, is rapidly elevated upon RTK activation within the framework of transcription-based negative feedback loops.

Signal Attenuation by Ligand-Induced Endocytosis of RTKs

Concomitant with receptor activation, ligand binding initiates a multi-step process that culminates in receptor degradation. RTKs such as EGFR and ErbB-2 are enriched in membrane microdomains called caveolae. This subset of lipid rafts contains Caveolin proteins, glycosphingolipids, and cholesterol, as well as multiple signaling molecules, including Src family kinases and H-Ras (reviewed in ref. 22). A conserved Caveolin-binding motif within the kinase domain of EGFR mediates the interaction of EGFR with the cytosolic Caveolin scaffolding domain of Caveolin -1 and -3.[23] Further, this interaction may inhibit kinase activity, but ligand binding to ErbB-1 induces migration of active receptors out of caveolae in a process requiring Src family kinases.[24] Subsequently, active RTK molecules aggregate over Clathrin-coated regions of the plasma membrane, where they start their journey to the lysosome, which will be detailed below. It is notable that RTK molecules whose internalization and degradation are defective due to large deletions acquire enhanced mitogenic and oncogenic activities.[25] This and similar observations are consistent with the notion that endocytosis serves to attenuate growth factor signals.

Common Molecular Mechanisms in Receptor Endocytosis: Curvature Sensing and Post-Translational Protein Modifications

Bending of the planar lipid bilayer and two post-translational protein modifications, namely phosphorylation and ubiquitylation, are involved in receptor endocytosis. These machineries, in combination with vesicle budding and fusion, recur along the endocytic itinerary, although distinct sets of protein platforms execute receptor sorting at different steps of endocytosis.

Membrane Bending

The generation of high-curvature lipid-bound transport carriers represented by tubules and vesicles requires physical perturbation of the lipid bilayer, as well as direct interactions between cytosolic proteins and lipid bilayers. Proteinaceous coats selectively associated with the surface of membrane buds are key mediators of vesicle formation in the endocytic pathways: Clathrin oligomerization into a coat scaffold on the membrane forms a polyhedral lattice. Nevertheless, the current notion is that Clathrin can at best serve to maintain an already curved membrane, thereby preventing its collapse back into a planar form.[26] Thus, in addition to coat-protein lattice formation mechanisms that help deforming the bilayer are likely to come into play. The GTPase called Dynamin was found to deform lipid bilayers into narrow tubules coated by Dynamin spirals.[27] Cytosolic proteins like Amphiphysin and Endophilin, two major interactors of Dynamin, were found to deform liposomes in vitro into narrow membrane tubules.[28,29] Both adaptors are able to sense and alter membrane curvature by modifying the lipid content, or by deforming the membrane mechanically via their BAR domains.[30] The BAR domain has a coiled-coil structure that binds preferentially to negatively charged membranes.[31] Epsin, an interactor of Clathrin and of the Clathrin adaptor AP-2, was also shown to induce membrane tubulation.[32] In addition, lipid components of the membrane, either directly or via interaction with proteins, have been suggested to facilitate the structural changes necessary to deform membranes. For example, selective transfer of lipids between bilayer leaflets has been proposed as a mechanism by which surface area asymmetries could influence budding and endocytosis. In addition, certain lipid species are postulated to favor bilayer curvature owing to their intrinsic properties, their relative geometries, or both.[33] Cholesterol, for example, selectively accumulates along the membrane, which may decrease membrane rigidity, create bilayer surface-area discrepancy, and facilitate budding.[34]

Protein Phosphorylation

Protein phosphorylation is a major regulatory mechanism of Clathrin-dependent endocytosis. Serine and threonine phosphorylation of Clathrin coat proteins plays an important role in the organization of macromolecular complexes during Clathrin-mediated endocytosis. Src-dependent tyrosine phosphorylation of coat proteins, such as Clathrin heavy chain and Dynamin, has been observed in cells stimulated with growth factors or hormones, although the precise role for these modifications is not fully understood.[35,36] Likewise, although Eps15 undergoes tyrosine phosphorylation, the precise role of this phosphorylation remains unclear. It appears, however, that phosphorylation-impaired mutants of Eps15 do not interfere with EGFR recruitment to pits, but rather with subsequent phases of the internalization process.[37] Phosphorylation of the hepatocyte growth factor-regulated tyrosine kinase substrate (Hrs; renamed Hgs) in response to EGF stimulation takes place on the evolutionary conserved tyrosines 329 and 334. Whereas, the ubiquitin-interacting motif (UIM; see below) of Hgs is required for phosphorylation, this modification is not required for UIM-dependent ubiquitylation.[38]

Protein Ubiquitylation

Modification of plasma membrane proteins by mono-ubiquitylation appears to serve as a signal sufficient to induce internalization and endocytic trafficking.[7,39] In line with this notion, two recent reports demonstrated that in-frame fusion of ubiquitin to EGFR resulted in constitutive internalization and enhanced degradation of the chimeric proteins.[40,41] Further, Cbl was shown to mediate mono-ubiquitylation of EGFR at multiple lysine residues, some localized within the kinase domain. These data demonstrate that mono-ubiquitylation is sufficient to induce endocytosis and lysososmal degradation of RTKs in mammalian cells, as has previously been reported in yeast.[42] Other studies performed with yeast cells demonstrated that in addition to cargo ubiquitylation, the endocytic machinery is also regulated by ubiquitylation.[43] At the plasma membrane of mammalian cells, machinery's components include Eps15 and Epsin, adaptors sharing an UIM, and like in yeast, the E3 ligase involved in their ubiquitylation is a member of the Rsp5p/Nedd4 family. The activity of Nedd4 family ligases may be regulated by RTKs; Nedd4 and the Nedd4-like ligase AIP4 are phosphorylated upon EGF stimulation.[44] Another regulatory mechanism has been uncovered in flies: the deubiquitylating enzyme Fat Facets was shown to de-ubiquitylate Liquid Facets, an orthologue of Epsin,[45] which is associated with endocytosis in *Drosophila*.[46] A mammalian homologue of Fat Facets, FAM, was localized to multiple points of E-cadherin and Beta-catenin trafficking,[47] but its association with RTK endocytosis remains to be elucidated.

The Journey of RTKs to the Lysosome

Receptor endocytosis involves several distinct steps and a continuously decreasing gradient of intravesicular pH. Internalization requires clustering of cargo molecules over Clathrin-coated regions of the cell surface, membrane bending and formation of a Clathrin-coated vesicular structure. Following its formation and pinching off, the shuttle vesicle loses its coat and fuses with an apparently stationary compartment, the early endosome. The next sorting event occurs in the multi-vesicular body (MVB). However, unlike cargo sorting at the plasma membrane, receptors delivered to the limiting membrane of the MVB are sorted 'away from the cytoplasm', into invaginations and internal vesicles, which accumulate lysosomal enzymes.

Figure 1 schematically delineates the multi-step sorting process and Table 1 lists the respective protein players. This highly coordinated but incompletely understood process entails many protein-lipid and protein-protein interactions, which are depicted in Figure 2, and its default route targets receptors back to the cell surface (recycling). Early studies that compared the routes of endocytosis of wild type EGFR and a kinase-defective mutant concluded that internalization at the plasma membrane is largely a kinase-independent process, which is followed by efficient recycling.[48,49] This pathway is similar to the route taken by Transferrin receptors, which are constitutively internalized even in the absence of a ligand. Clathrin, along with its

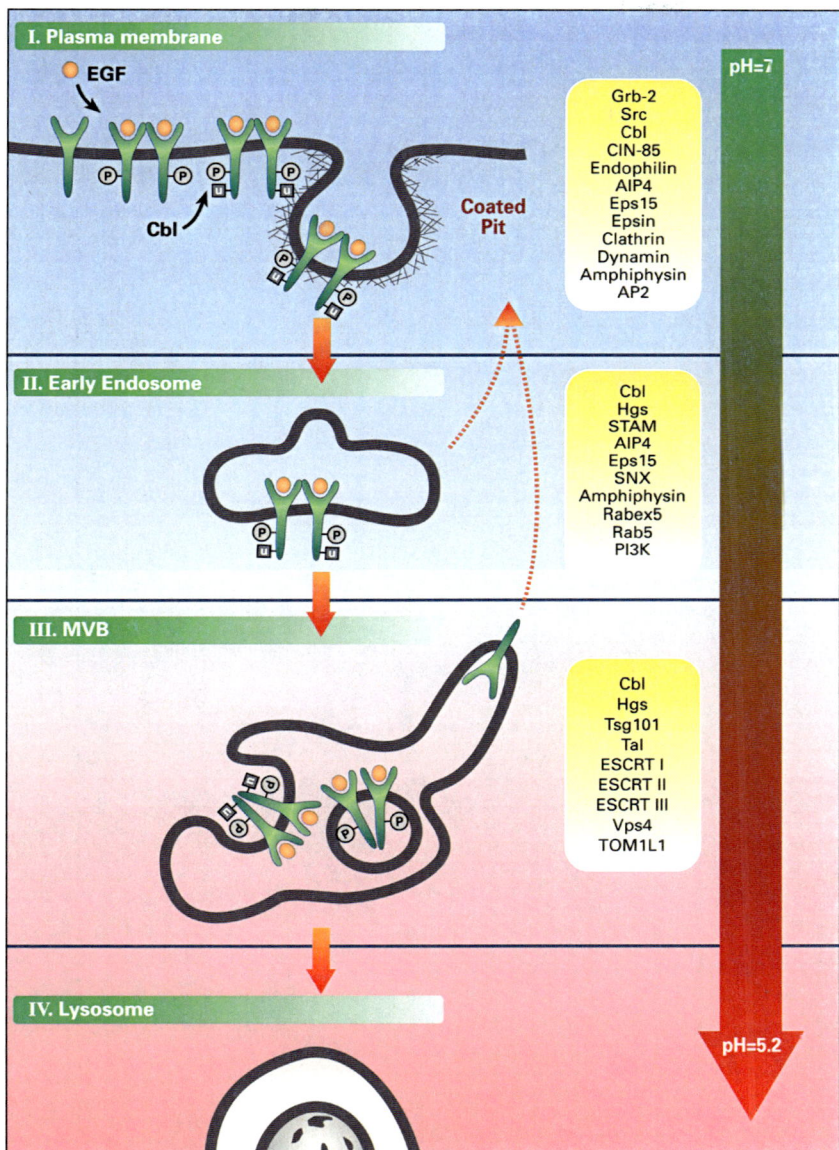

Figure 1. The journey of RTKs to the lysosome. Upon ligand binding at the plasma membrane, RTK molecules like EGFR cluster over Clathrin-coated pits, which subsequently pinch-off to form Clathrin-coated vesicles. The vesicle looses the Clathrin coat and fuses with the early endosome. Next, receptors are sorted to internal vesicles of the MVB and subsequently to lysosomes, where they undergo degradation by hydrolases. Alternatively, RTK molecules recycle back to the plasma membrane (dotted arrow). The major regulators of each sorting compartment are listed in the respective box, and the decreasing gradient of intravesicular pH is indicated by the vertical arrow. P letters refer to tyrosine phopshorylation and U letters refer to mono-ubiquitylation.

Table 1. Proteins involved in endocytosis of RTKs

Site	Species		Domains	Role
	Mammals	**Yeast**		
Membrane	EGFR	-	L/CR/JM/TKD/CT	RTK
	Src	-	Unique/SH3/SH2/TKD	Phosphorylates CHC and redistributes clathrin upon EGFR activation
	c-Cbl	-	TKB/RF/Pro/UBA	Ubiquitin E3 ligase
	AIP4	Rsp5p	C2/WW/HECT	Ubiquitin E3 ligase; interacts with Endophilin, Epsin, Eps15 and Cbl
	Epsin	Ent1/Ent2	ENTH/UIM/CBM/AEBM/NPF	Clathrin coat adaptor; AP2 binding protein
	Eps15	Ede1	EH/CC/UIM	Clathrin coat adaptor
	Endophilin	Rvs167	SH3/hinge region/NT/CC	Interacts with dynamin, AIP4 and CIN85; lipid modifying enzyme(LPAAT)
	CIN-85/SETA	Sla-1	SH3/Pro/CC	Adaptor protein; interacts with Cbl and endophilin; regulates clathrin mediated endocytosis
Early endosome	Amphiphysin	Rvs161p/Rvs167p	BAR/Pro/CLAP/SH3	Recruits dynamin; interacts with AP2
	Dynamin	Dynamin	GTPase/middle/PH/GED/PRD	Forms collars on invaginating pits
	Hgs	Vps27p	VHS/FYVE/UIM/Pro/CC/PG	Regulates endosomal sorting
	STAM	Hse-1	VHS/UIM/SH3/ITAM	Interacts with Hgs
	SNX	Vps5p/Vps17p/YIL036W	PX/CC	Regulates intracellular sorting; interacts with Hgs and amphiphysin
MVB	Rabex	Vps9	GEF/CUE	Activates Rab5 and cooperates with other factors to promote endosome fusion
	Tal	-	LRR/ERM/CC/SAM/PTAP-PSAP/RF	Ubiquitylates Tsg101
	TOM1L1	ND	VHS/PTAP/GAT/GBD/SH3-BD/PBD/SH2-BD	Interacts with Hgs and Tsg101
	Tsg101	Vps23p	UEV/Pro/CC/SB	Recruits ubiquitylated proteins to ESCRTI

Table continued on next page

Table 1. *Continued*

Site	Species		Domains	Role
	Mammals	**Yeast**		
MVB	ESCRTI:			Involved in the sorting of ubiquitylated proteins
	Tsg101/	Vps23/	UEV/CC	
	Vps28/	Vps28/		
	Vps37c	Vps37	CC	
	ESCRTII:			Transiently recruited to endosomal membranes; required recruitment of ESCRTIII
	EAP25/	Vps25/	CC	
	EAP30/	Vps22/	CC	
	EAP45	Vps36	NZF	
	ESCRTIII:			Polymerize to form intraluminal vesicles
	CHMP2/	Vps2/	CC	
	CHMP6/	Vps20/	CC	
	CHMP3/	Vps24/	CC	
	CHMP4	Vps32	CC	
	VPS4A,B	Vps4	CC/AAA	Dissociates ESCRT from the membrane de-ubiquitylating enzyme
	UBP	Doa4	CC/RD/CB/HB	

Protein regulators of endocytic sorting of RTKs like EGFR are listed according to their site of action. Note that the ubiquitin-specific processing protease (UBP) refers to UBPY/UBP8, which belongs to a family of mammalian ubiquitin hydrolases related to yeast Doa4p. Ubiquitin-binding proteins like Eps15, Epsin, Hgs and STAM are involved in ubiquitin-dependent regulation of receptor trafficking. The abbreviations used are: AEBM: alpha-ear binding motif; BAR: Bin-Amphiphysin-Rvsp; CB: cysteine box; CBM: Clathrin-binding motif; CC: : coiled coil; CLAP: Clathrin-AP2 binding; CR: cysteine-rich; CT: carboxyl-terminal; CUE: Cue1p-homology; EH: Eps15 homology; ENTH: Epsin N-terminal homology; ERM: Ezrin-Radixin-Moesin; FYVE: Fab1/YOTB/Vac1/EEA1; GAT: GGA and Tom-1; GBD: Grb2 binding domain; GED: GTPase effector domain; GEF: guanine-nucleotide exchange factor; HB: histidine box; HECT: homologous to E6-AP C-terminus; ITAM: immunoreceptor tyrosine-based activation motif; JM: juxtamembrane; L: ligand binding; LPAAT: lysophosphatidic acid acyltransferase; LRR: leucine-rich repeats; NT: N-terminal alpha helical; PBD: p65 binding domain; PG: proline- and glutamine-rich; PH: pleckstrin homology; PRD: proline- and arginine-rich domain; Pro: proline-rich; PX: Phox; RF: ring finger; RD: rhodanese-homology domain; SAM: sterile alpha motif; SB: steadiness box; SH2: Src homology 2; SH2-BD: SI2 binding domain; SH3: Src homology 3; SI I3-BD: SH3 binding domain; 1kB: tyrosine kinase binding; TKD: tyrosine kinase binding; UBA: ubiquitin associated; UEV: ubiquitin E2 variant; UIM: ubiquitin interaction motif; Unique: a region which varies among family members; VHS: Vps27p/Hrs/STAM.

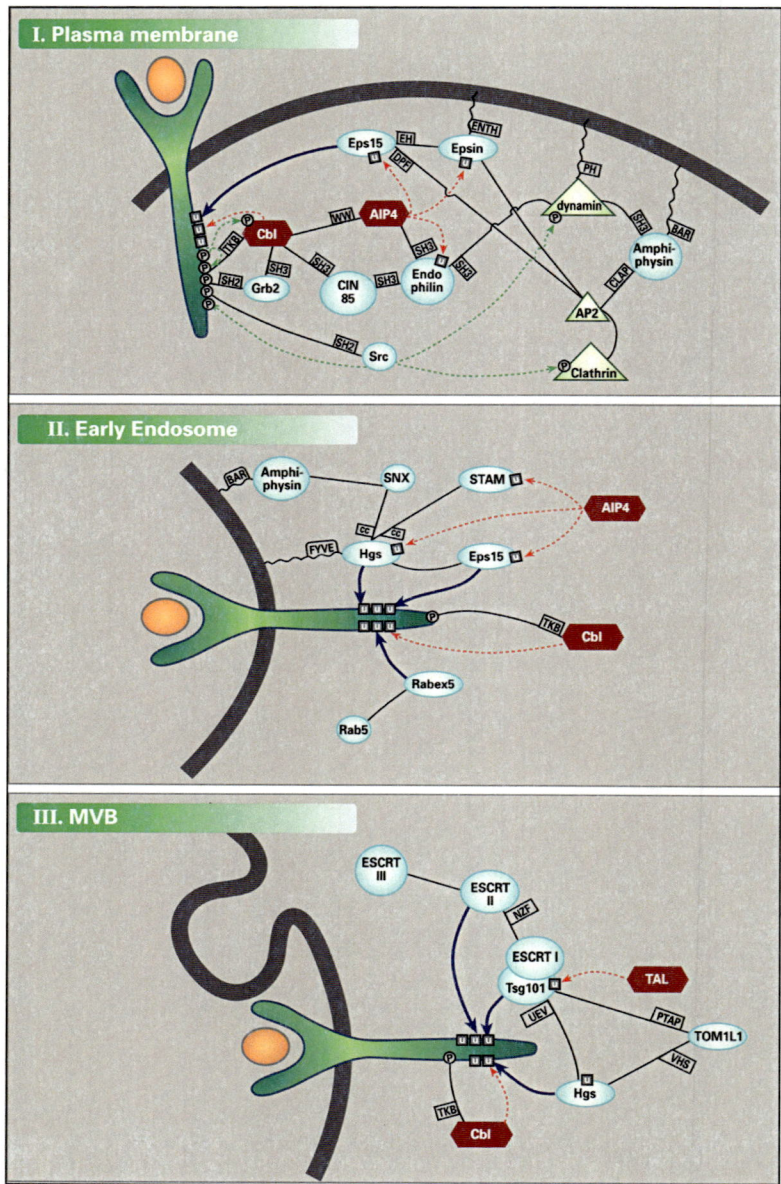

Figure 2. Protein networks involved in endocytosis of RTKs. Protein-protein and protein-lipid interactions, as well as phosphorylation and ubiquitylation of protein substrates, are illustrated for each step in the journey of an RTK like EGFR to the lysosome. Bold arrows (blue) indicate ubiquitin-mediated protein-protein interactions (ubiquitin-binding proteins are shown in ellipses). Dotted arrows refer to protein modifications, either phopshorylation (P; green) or ubiquitylation (U: red). E3 ligases are shown as hexagons. Some of the protein motifs involved in protein-protein interactions and protein-lipid binding (wavy lines) are shown in flags.

adaptor, AP2, and the GTPase Dynamin, play pivotal roles in the constitutive pathway of RTK endocytosis. To escape the recycling route, RTKs must recruit ubiquitin ligases, along with a group of mono-ubiquitin binding proteins (e.g., Eps15 and Epsin), which enable selective sorting of mono-ubiquitylated receptors, primarily at the early endosome and at the MVB.

Receptor Sorting at the Plasma Membrane

In the absence of ligands, a large fraction (40-60%) of EGFRs is found in caveolae, which are enriched in several signaling proteins, including Src. Random exit from the caveolae may be enhanced upon trans- or auto-phosphorylation of EGFR, primarily on tyrosine 845, recruitment and activation of Src.[50] Subsequently, Src phosphorylates both Dynamin[35] and the heavy chain of Clathrin, which is followed by Clathrin redistribution to the cell periphery[36] and enhanced endocytosis of EGFR.

The Clathrin-AP2-Amphiphysin Platform

Although the polyhedral lattice of Clathrin is the main component of budding vesicles, it offers only a mechanical scaffold incapable of selecting cargo. Sorting is carried out by several adaptors, the major one at the plasma membrane being the heterotetrameric complex AP2. AP2 binds specific cytoplasmic motifs of transmembrane cargoes, thus allowing their inclusion into coated pits. For example, in the case of EGFR, AP2 binds truncated receptor mutants through the motif [974]YRAL,[51] but alternative mechanisms seem to mediate endocytosis in the context of full-length receptors.[52] Another important function of AP2 is recruitment of Amphiphysins, ligands of Dynamin, which are anchored at the membrane through a BAR (Bin-Amphiphysin-Rvsp) domain. This domain senses, and in the case of Amphiphysin also imposes membrane curvature.[31] Further, during the budding process, Amphiphysin recruits Dynamin to the neck of the nascent vesicle, where Dynamin forms a ring-like collar that finalizes the fission process.

The Cargo Ubiquitylation Platform: Grb2-Cbl-CIN85

Ligand-induced activation of several RTKs leads to direct or indirect recruitment and phosphorylation of Cbl proteins,[53-55] whose orthologue in worms is a major negative regulator of EGFR signaling.[56] Cbl is an E3 ubiquitin ligase that binds tyrosine phosphorylated cargoes like EGFR through an amino-terminal SH2-like domain and recruits an ubiquitin-loaded E2 molecule to the RING domain. While it is clear that c-Cbl is recruited to the plasma membrane upon activation of EGFR[57-61] and internalization of EGFR is reduced when the interaction with c-Cbl is interrupted,[62,63] several reports suggest that receptor endocytosis may take place in the absence of an intact ubiquitylation machinery.[64] In addition to the E3 ligase activity, c-Cbl recruits a large number of protein partners, including Grb2 and CIN85. Both direct and indirect (via Grb2) recruitment of c-Cbl leads to receptor ubiquitylation.[63] However, these alternative modes may not be functionally redundant; Grb2 seems essential for receptor endocytosis[65] and its knockdown indicates an essential function in the initial steps of EGFR internalization.[59] This may reflect the ability of Grb2 to interact with CIN85, a scaffold molecule and an ubiquitylation substrate that recruits Endophilins to internalizing RTKs.[66,67] The SH3 domain of the Endophilin adaptor binds Dynamin, whereas the N-terminal region possesses a lysophosphatidic acid acyltransferase (LPAAT) activity, which may assist in curving the planar plasma membrane.

The Coat-Adaptor Ubiquitylation Platform: AIP4-Eps15-Epsin

Evidence in yeast indicates that ubiquitylation of components other than the cargo is required for receptor endocytosis: the E3 ligase Rsp5p ubiquitylates a component of the endocytic machinery prior to endocytosis of the membrane proteins Gap1 and Ste2p.[43,68] The mammalian orthologues of Rsp5p, AIP4 and Nedd4, mediate ubiquitylation of several coat adaptors sharing an ubiquitin-interacting motif (UIM). At the plasma membrane, these include Eps15 and Epsins.[69-74] Eps15 includes multiple protein-protein interaction modules; three Eps15 homology (EH) domains, which bind the NPF tripeptide motif, a homodimerization coiled-coil domain, multiple DPF tripeptide motifs, which bind to the Clathrin coat via AP2, and a tandem UIM. In addition to the UIM, Epsin harbors a $PI(4,5)P_2$ binding domain, the Epsin N-terminal homology (ENTH) domain, which plays a role in the initiation of the budding process.[32] Epsins are recruited to biological membranes by several additional interactions, which involve Clathrin, AP2 and Eps15. Upon EGF stimulation, the latter undergoes tyrosine phosphorylation, recruitment to the plasma membrane[75] and mono-ubiquitylation.[76] Several models have recently been proposed for how ubiquitin-binding motifs direct mono-ubiquitylation of coat-adaptors like Eps15. Accordingly, the UIM may bind ubiquitin in the thiol-ester intermediate state of E3 ubiquitin ligases.[77] Ubiquitin is then transferred to the UIM-containing protein, which no longer interacts with the E3 ligase and is not subject to poly-ubiquitylation. According to an alternative model, intramolecular interactions between the UIM and ubiquitin masks the lysine residue at position 48, the main site for ubiquitin branching, thus inhibiting further chain assembly.[78]

Receptor Sorting at the Early Endosome

Once sorted to Clathrin-coated vesicles, internalized receptors are delivered within 2-5 minutes to a tubular-vesicular network located in the cell periphery. The Clathrin-coated vesicle sheds Clathrin and fuses with an internal vesicle to form the early endosome. Endocytic vesicle maturation is concomitant with a reduction in the internal pH and accumulation of hydrolytic enzymes. This endosomal trafficking is controlled by a group of GTPases, primarily Rab5, which are regulated by EGFR. The early endosome is a major site for sorting of endocytosed cell surface receptors, which are to be recycled to the cell surface or destined for degradation in lysosomes (reviewed by ref. 79). Newly synthesized lysosomal hydrolases are also delivered to the lysosome from the trans-Golgi network via the endosome. At this organelle, ubiquitylation serves as a sorting signal for endocytosed receptors and newly synthesized lysosomal proteins to be incorporated into the luminal vesicles of the MVB.[80] Presumably, ubiquitin-mediated sorting of cargoes at the endosome shares some attributes with the process which occurs at the membrane, but a partly different set of UIM-containing adaptors and their direct partners, namely: Hrs/Hgs, STAM/East, Hbp and Sorting nexins (SNX), participate in endosomal sorting.[81]

The Role for Rab5

The Rab5 small GTPase plays a central role in the formation of endosomes by regulating Clathrin-coated vesicle formation, the fusion of endocytic vesicles, and their movements along microtubules (reviewed in ref. 82). Several studies have established that the activated EGFR modulates the GTPase activity of Rab5 by targeting either GTPase-activating proteins (GAPs), such as RN-Tre, or GTP exchange factors (GEFs), such as Rabex-5/Vps9p or RIN1.[83] A critical effector of Rab5 is Rabaptin-5, which forms a complex with Rabex-5. Upon activation of Rab5 by Rabex-5, the Rabaptin-Rabex complex induces its own membrane recruitment through Rabaptin-5. This positive feedback loop is thought to create a microenvironment enriched in active Rab5 on the membrane, where other Rab5 effectors are concentrated. The activation of Rab5 through EGFR and RIN1 stimulates both the internalization and degradation of activated EGFRs, as also occurs upon expression of constitutively active mutants of Rab5 (reviewed in ref. 84). Recently, two new protein partners for Rab5, APPL1 and APPL2, were discovered.[85] These proteins bind to the active, GTP-bound form of Rab5 and label a sub-population of peripheral endosomes to which a fraction of the internalized EGFR is targeted.

The Hgs-STAM-SNX Platform

Hrs/Hgs was originally characterized as a tyrosine phosphorylation substrate for several growth factor receptors, including EGFR. In addition to a single UIM, Hgs is composed of several recognizable domains: an amino-terminal VHS domain, the FYVE phosphatidylinositol 3-phosphate binding domain responsible for endosomal localization,[86] a proline-rich region, and a coiled-coil domain, which recruits Sorting Nexin 1 (SNX-1). Apart from the phox-homology (PX) domain, through which SNX-1 binds PI3P and PI(3,5)P$_2$, SNX-1 contains a BAR domain, which endows the ability to form dimers and sense the high membranes curvature of early endosomes.[87] Similarly, SNX-16 directs sorting of EGFR to the endosomal compartment and regulates EGF-induced signaling.[88] Other ligands of the coiled-coil domain of Hgs are the signal transducing adaptor molecule (STAM/East) and the Hgs-binding protein (Hbp), both implicated in the regulation of growth factor receptor levels and signaling.[89,90] Through interactions with ubiquitylated cargo proteins on the early endosome via the VHS and UIM domains, STAM participates in the sorting of cargo proteins for trafficking to the lysosome.[91] Apparently, Hgs forms a multivalent complex with STAM and Eps15. The localization of this complex to Clathrin-rich regions of the endosome membrane is controlled by Vps4, a AAA-type ATPase, which has been implicated in MVB formation.[92]

The Tom-1 Platform

Tom-1 (target of Myb1) was identified as a protein whose expression is induced by a viral oncoprotein (v-Myb). The conserved amino-terminal domains of Tom-1 and its relative, Tom-1L1 (also referred to as Srcasm;[93]), harbor a VHS (Vps27p/Hrs/STAM) domain followed by a GAT (GGA and Tom-1) domain. The GAT domain of Tom-1 binds ubiquitin and Tollip (Toll-interacting protein) in a mutually exclusive manner. Interestingly, Tollip recruits Tom-1 and ubiquitylated proteins to the early endosome.[94] Thus like Hrs, Tom-1 is involved in sorting of ubiquitylated proteins into clathrin-coated microdoamins of early endosomes and the MVB.

Receptor Sorting at the MVB

During their maturation, early endosomes loose their tubular extensions, translocate along microtubules toward the nucleus and become more acidic. This process leads to the formation of the late endosome, a vesicular compartment that receives no direct transport of vesicles from the plasma membrane. Late endosomes are dynamic compartments with pleiomorphic organization, containing cisternal, tubular and vesicular regions with numerous membrane invaginations, which gave them the name multivesicular bodies (MVBs). This prelysosomal organelle is enriched in proteins targeted for degradation, and its limiting membrane contains high amounts of LAMP1, a characteristic protein of lysosomes. Fusion of the limiting membrane of the MVB with the lysosomal membrane results in the delivery of luminal vesicles and their contents to the hydrolytic interior of lysosomes, where they are degraded. Ubiquitin is thought to act as a positive signal for cargo sorting in the MVB. The corresponding components of the MVB sorting machinery were uncovered in large part by genetic analyses in yeast, which led to the identification of many proteins implicated in vacuolar protein sorting (called Vps; reviewed in ref. 95). Sorting of ubiquitylated proteins into the MVB pathway is executed by three distinct complexes of class E Vps proteins, called endosomal complexes required for transport (ESCRT-I, -II, and -III), and the AAA-type ATPase Vps4.

The ESCRT-I Platform

In yeast, the 350kDa ESCRT-I complex is composed of Vps23, Vps28 and Vps37, and it plays a crucial role in the selection of ubiquitylated cargoes.[96] The cargo-binding component is most likely Vps23p, which harbors a catalytically inactive ubiquitin conjugating-like (UBC-like) domain, also called ubiquitin E2 variant (UEV) domain. The mammalian orthologue of Vps23p, tumor susceptibility gene 101 (Tsg101), was originally isolated by using an anti-sense RNA screen for malignant transformation of murine fibroblasts.[97] Tsg101 has been shown to bind ubiquitin through the UEV domain, which includes a second binding site for a tetrad amino acid motif, P(T/S)AP.[98] When Tsg101 is ablated, internalized EGFR molecules are shunted from the degradative pathway to a recycling route that enhances and prolongs signaling.[99] This may contribute to the tumorigenic phenotype exhibited by fibroblasts in which tsg101 expression has been ablated. Tsg101 directly interacts with Hgs via a PSAP motif, which binds the

UEV domain of Tsg101.[100] Hgs-Tsg101 interactions are required for EGFR transport from the early to the late endosomes. Presumably, Hgs binds ubiquitylated cargo and recruits Tsg101, which then also interacts with ubiquitin moieties on the cargo and assembles ESCRT complexes for subsequent trafficking.[101] Tsg101 undergoes mono-ubiquitylation by an E3 ligase called Tsg101-associated ligase (Tal).[102] Upon ubiquitylation, Tsg101 is no longer capable of EGFR sorting, presumably because its own ubiquitins block cargo loading at the UEV. Thus cyclic ubiquitylation and de-ubiquitylation of Tsg101 by Tal and an unknown deubiquitylation enzyme may underlie ESCRT-1-mediated sorting of ubiquitylated cargoes into the lumen of the MVB in animal cells.

The ESCRT-II and ESCRT-III Platforms

The 155 kDa soluble ESCRT-II complex of yeast includes Vps22, Vps25, and Vps36. This protein complex transiently associates with the endosomal membrane and thereby initiates the formation of ESCRT-III. Like other sorting platforms, ESCRT-II contains an ubiquitin-binding subunit, Vps36, which binds ubiquitin via an NZF (Np14 zinc finger) domain.[103] In analogy to ESCRT-I, ESCRT-II selects and sort MVB cargoes for delivery to the lumen of the lysosome.[104] Removal of ubiquitin from MVB cargoes, which occurs in yeast cells before the cargo enters the luminal vesicles of an MVB, requires the enzymatic activity of a de-ubiquitylating enzyme called Doa4 (degradation of alpha-2), a homologue of certain mammalian ubiquitin hydrolases (e.g., UBPY). Doa4 recruitment to the endosome requires the correct assembly of ESCRT-III.[105] After ubiquitin removal, cargoes are sorted into invaginating vesicles that eventually bud into the lumen. This requires the function of additional regulators, which are currently unknown (reviewed in ref. 95). Finally, the disassembly and release of the entire MVB sorting machinery, which allows the ESCRT machinery to recycle back into the cytoplasm, is controlled by the AAA-type ATPase Vps4. Future studies will likely uncover the role of deubiquitylation enzymes and other mammalian proteins associated with MVB sorting. One interesting candidate is Tom-1L1.[106] The VHS domain of Tom-1L1 interacts with Hgs, while its PTAP motif is responsible for binding to Tsg101. In addition, Tom-1L1 possesses several tyrosine motifs at the C-terminal region that mediate interactions with members of Src family kinases and other signaling proteins, such as Grb2 and p85-PI3K. Thus Tom-1L1 recruitment to the sorting machinery of MVB may induce activation of signaling complexes.

The Interface of Receptor Trafficking and Signaling

In general, ligand-induced receptor endocytosis serves as a machinery that terminates growth factor signaling. Consistent with this notion, SLI-1, the single Cbl orthologue of *C. elegans*, is a major negative regulator of EGFR signaling,[107] and an internalization-defective mutant of EGFR is characterized by enhanced signaling.[63] The default recycling pathway enables prolonged signaling, which explains the oncogenic action of transforming mutants of c-Cbl.[62] The efficiency of recycling decreases as receptors reach late compartments of endocytosis. Nevertheless, this process, unlike sorting for degradation, does not require the intrinsic kinase activity.[108] The mechanisms affecting recycling are incompletely characterized. For example, threonine phosphorylation of EGFR by protein kinase C inhibits receptor ubiquitylation and enhances recycling,[109] but the underlying mechanism remains unknown. In the case of G protein-coupled receptors there are at least two recycling pathways, a direct (fast) pathway that depends on Rab4, and a slow recycling route mediated by both Rab4 and Rab5 (reviewed in ref. 110), but RTK recycling is less characterized. Nevertheless, it has been reported that Rab11 plays a role in late recycling of EGFR.[111] Accumulating evidence indicates that the internalized EGFR continues to bind and phosphorylate downstream signaling proteins in predegradative intracellular compartments, leading to activation of signaling pathways distinct from those originated at the cell surface (reviewed in ref. 112). Internalized EGFRs are enzymatically active, hyperphosphorylated and associated with Ras-GAP, Shc, Grb2 and mSOS.[113] Moreover, endosomal EGFR signaling is sufficient to activate the major signaling pathways leading to cell proliferation and survival, as well as suppression of apoptosis induced by serum withdrawal.[114]

Ligand-Independent Pathways of RTK Endocytosis

Accumulating evidence show that in addition to the well-characterized ligand-induced pathway of RTK down-regulation, there are alternative, ligand-independent mechanisms for receptor internalization and degradation. For example, anti-receptor antibodies cause relatively slow endocytosis and degradation of EGFR and ErbB-2/HER2.[115] Because certain monoclonal antibodies to these RTKs are clinically used to treat various types of cancer (reviewed in ref. 116), the mechanism of antibody-mediated endocytosis of RTKs is relevant to therapeutic applications. Due to their bivalence, most anti-receptor antibodies weakly activate tyrosine auto-phosphorylation and downstream signaling, including phosphorylation of c-Cbl. This may account for the ability of antibodies to elevate receptor ubiquitylation.[117] Nevertheless, antibodies can down regulate RTKs through a c-Cbl- and ubiquitylation-independent mechanism, whose rate is proportional to the size of antibody-receptor lattices formed at the cell surface.[118] Unlike anti-receptor antibodies, which interact directly with the internalizing receptor, agonists of G protein-coupled receptors, such as the beta$_2$-adrenergic receptor, indirectly act upon EGFR. These agonists induce dimerization and auto-phosphorylation of EGFR, followed by beta-arrestin-dependent internalization of both EGFR and the beta$_2$-adrenergic receptor.[119] Yet a third mechanism of ligand-independent endocytosis of RTKs is put into motion under stress conditions. These include osmotic stress, stimulation with the tumor necrosis factor and irradiation by ultraviolet (UV) light. Immunofluorescence microscopy demonstrated that upon UV treatment of cells, EGFR translocates to internal vesicles but undergoes no tyrosine phosphorylation, ubiquitylation or degradation.[120-122] On the other hand, oxidative stress induces extensive tyrosine phosphorylation of EGFR, but the phosphorylated receptor fails to recruit c-Cbl and the ubiquitylation machinery.[123] Clearly, ligand-mediated endocytosis of RTKs is not the only way to internalize active receptors and the intracellular routing is dictated by the type of stimulus. However, the mechanisms underlying the alternative routes of RTK endocytosis remain largely obscure.

Perspectives

Owing to the inherent dynamics and complexity of the endosomal system, our understanding of endocytosis and the signals that target receptor molecules to different intracellular pathways is still incomplete. The importance of in-depth understanding stems from the oncogenic potential of many members of the RTK family and the association of transforming ability with localization at the cell surface. Nevertheless, recent analyses of EGFR and several other receptors, such as the nerve growth factor receptor and the transforming growth factor-beta receptor, indicate that receptors stimulate distinct signaling pathways when residing in different sub-cellular compartments. Along with exhaustive identification of the respective molecular players and their post-translational modifications, sophisticated microscopic methods will likely contribute to future efforts to harness or manipulate endocytic pathways for clinical applications.

Acknowledgements

We thank members of our group for useful insights. Our laboratory is supported by research grants from Minerva, the Prostate Cancer Foundation, the National Cancer Institute (grant CA72981), the M.D. Moross Institute for Cancer Research, The Israel Science Foundation and the Willner Family Center for Vascular Biology. Y.Y. is the incumbent of the Harold and Zelda Goldenberg Professorial Chair.

References

1. Schlessinger J. Cell signaling by receptor tyrosine kinases. Cell 2000; 103:211-225.
2. Yarden Y, Ullrich A. Growth factor receptor tyrosine kinases. Ann Rev Biochem 1988; 57:443-478.
3. Blume-Jensen P, Hunter T. Oncogenic kinase signaling. Nature 2001; 411:355-365.
4. Pawson T. Specificity in signal transduction: From phosphotyrosine-SH2 domain interactions to complex cellular systems. Cell 2004; 116:191-203.

5. Dikic I, Giordano S. Negative receptor signalling. Curr Opin Cell Biol 2003; 15:128-135.
6. Polo S, Pece S, Di Fiore PP. Endocytosis and cancer. Curr Opin Cell Biol 2004; 16:156-161.
7. Marmor MD, Yarden Y. Role of protein ubiquitylation in regulating endocytosis of receptor tyrosine kinases. Oncogene 2004; 23:2057-2070.
8. Burke P, Schooler K, Wiley HS. Regulation of epidermal growth factor receptor signaling by endocytosis and intracellular trafficking. Mol Biol Cell 2001; 12:1897-1910.
9. Lowenstein EJ, Daly RJ, Batzer AG et al. The SH2 and SH3 domain-containing protein GRB2 links receptor tyrosine kinases to ras signaling. Cell 1992; 70:431-442.
10. Pelicci G, Lanfrancone L, Grignani F et al. A novel transforming protein (SHC) with an SH2 domain is implicated in mitogenic signal transduction. Cell 1992; 70:93-104.
11. Seger R, Krebs EG. The MAP kinase signaling cascade. FASEB J 1995; 9:726-735.
12. Soltoff SP, Carraway IIIrd KL, Prigent SA et al. ErbB3 is involved in activation of phosphatidylinositol 3-kinase by epidermal growth factor. Mol Cell Biol 1994; 14:3550-3558.
13. Cantley LC, Neel BG. New insights into tumor suppression: PTEN suppresses tumor formation by restraining the phosphoinositide 3-kinase/AKT pathway. Proc Natl Acad Sci USA 1999; 96:4240-4245.
14. Rhee SG. Regulation of phosphoinositide-specific phospholipase C. Ann Rev Biochem 2001; 70:281-312.
15. Moghal N, Sternberg PW. Multiple positive and negative regulators of signaling by the EGF-receptor. Curr Opin Cell Biol 1999; 11:190-196.
16. Shilo BZ. Signaling by the Drosophila epidermal growth factor receptor pathway during development. Exp Cell Res 2003; 284:140-149.
17. Hackel PO, Gishizky M, Ullrich A. Mig-6 is a negative regulator of the epidermal growth factor receptor signal. Biol Chem 2001; 382:1649-1662.
18. Fiorentino L, Pertica C, Fiorini M et al. Inhibition of ErbB-2 mitogenic and transforming activity by RALT, a mitogen-induced signal transducer which binds to the ErbB-2 kinase domain. Mol Cell Biol 2000; 20:7735-7750.
19. Azios NG, Romero FJ, Denton MC et al. Expression of herstatin, an autoinhibitor of HER-2/neu, inhibits transactivation of HER-3 by HER-2 and blocks EGF activation of the EGF receptor. Oncogene 2001; 20:5199-5209.
20. Tsang M, Friesel R, Kudoh T et al. Identification of Sef, a novel modulator of FGF signalling. Nat Cell Biol 2002; 4:165-169.
21. Sun H, Charles CH, Lau LF et al. MKP-1 (3CH134), an immediate early gene product, is a dual specificity phosphatase that dephosphorylates MAP kinase in vivo. Cell 1993; 75:487-493.
22. Anderson RG. The caveolae membrane system. Annu Rev Biochem 1998; 67:199-225.
23. Couet J, Sargiacomo M, Lisanti MP. Interaction of a receptor tyrosine kinase, EGF-R, with caveolins. Caveolin binding negatively regulates tyrosine and serine/threonine kinase activities. J Biol Chem 1997; 272:30429-30438.
24. Mineo C, Gill GN, Anderson RG. Regulated migration of epidermal growth factor receptor from caveolae. J Biol Chem 1999; 274:30636-30643.
25. Wells A, Welsh JB, Lazar CS et al. Ligand-induced transformation by a noninternalizing epidermal growth factor receptor. Science 1990; 247:962-964.
26. Nossal R. Energetics of clathrin basket assembly. Traffic 2001; 2:138-147.
27. Marks B, Stowell MH, Vallis Y et al. GTPase activity of dynamin and resulting conformation change are essential for endocytosis. Nature 2001; 410:231-235.
28. Takei K, Slepnev VI, Haucke V et al. Functional partnership between amphiphysin and dynamin in clathrin-mediated endocytosis. Nat Cell Biol 1999; 1:33-39.
29. Farsad K, Ringstad N, Takei K et al. Generation of high curvature membranes mediated by direct endophilin bilayer interactions. J Cell Biol 2001; 155:193-200.
30. Zimmerberg J, McLaughlin S. Membrane curvature: How BAR domains bend bilayers. Curr Biol 2004; 14:R250-252.
31. Peter BJ, Kent HM, Mills IG et al. BAR domains as sensors of membrane curvature: The amphiphysin BAR structure. Science 2004; 303:495-499.
32. Ford MG, Mills IG, Peter BJ et al. Curvature of clathrin-coated pits driven by epsin. Nature 2002; 419:361-366.
33. Burger KN. Greasing membrane fusion and fission machineries. Traffic 2000; 1:605-613.
34. Farsad K, De Camilli P. Mechanisms of membrane deformation. Curr Opin Cell Biol 2003; 15:372-381.
35. Ahn S, Kim J, Lucaveche CL et al. Src-dependent tyrosine phosphorylation regulates dynamin self-assembly and ligand-induced endocytosis of the epidermal growth factor receptor. J Biol Chem 2002; 277:26642-26651.

36. Wilde A, Beattie EC, Lem L et al. EGF receptor signaling stimulates Src kinase phosphorylation of clathrin, influencing clathrin redistribution and EGF uptake. Cell 1999; 96:677-687.
37. Confalonieri S, Salcini AE, Puri C et al. Tyrosine phosphorylation of Eps15 is required for ligand-regulated, but not constitutive, endocytosis. J Cell Biol 2000; 150:905-912.
38. Urbe S, Sachse M, Row PE et al. The UIM domain of Hrs couples receptor sorting to vesicle formation. J Cell Sci 2003; 116:4169-4179.
39. Polo S, Confalonieri S, Salcini AE et al. EH and UIM: Endocytosis and more. Sci STKE 2004; 213:re17.
40. Haglund K, Sigismund S, Polo S et al. Multiple monoubiquitination of RTKs is sufficient for their endocytosis and degradation. Nat Cell Biol 2003; 5:461-466.
41. Mosesson Y, Shtiegman K, Katz M et al. Endocytosis of receptor tyrosine kinases is driven by monoubiquitylation, not polyubiquitylation. J Biol Chem 2003; 278:21323-21326.
42. Hicke L, Riezman H. Ubiquitination of a yeast plasma membrane receptor signals its ligand-stimulated endocytosis. Cell 1996; 84:277-287.
43. Dunn R, Hicke L. Multiple roles for Rsp5p-dependent ubiquitination at the internalization step of endocytosis. J Biol Chem 2001; 276:25974-25981.
44. Courbard JR, Fiore F, Adelaide J et al. Interaction between two ubiquitin-protein isopeptide ligases of different classes, CBLC and AIP4/ITCH. J Biol Chem 2002; 277:45267-45275.
45. Chen X, Zhang B, Fischer JA. A specific protein substrate for a deubiquitinating enzyme: Liquid facets is the substrate of Fat facets. Genes Dev 2002; 16:289-294.
46. Cadavid AL, Ginzel A, Fischer JA. The function of the Drosophila fat facets deubiquitinating enzyme in limiting photoreceptor cell number is intimately associated with endocytosis. Development 2000; 127:1727-1736.
47. Murray RZ, Jolly LA, Wood SA. The FAM deubiquitylating enzyme localizes to multiple points of protein trafficking in epithelia, where it associates with E-cadherin and beta-catenin. Mol Biol Cell 2004; 15:1591-1599.
48. Opresko LK, Chang CP, Will BH et al. Endocytosis and lysosomal targeting of epidermal growth factor receptors are mediated by distinct sequences independent of the tyrosine kinase domain. J Biol Chem 1995; 270:4325-4333.
49. Honegger AM, Dull JT, Felder S et al. Point mutation at the ATP binding site of EGF receptor abolishes protein-tyrosine kinase activity and alters cellular routing. Cell 1987; 51:199-209.
50. Tice DA, Biscardi JS, Nickles AL et al. Mechanism of biological synergy between cellular Src and epidermal growth factor receptor. Proc Natl Acad Sci USA 1999; 96:1415-1420.
51. Sorkin A, Mazzotti M, Sorkina T et al. Epidermal growth factor receptor interaction with clathrin adaptors is mediated by the Tyr974-containing internalization motif. J Biol Chem 1996; 271:13377-13384.
52. Nesterov A, Carter RE, Sorkina T et al. Inhibition of the receptor-binding function of clathrin adaptor protein AP-2 by dominant-negative mutant mu2 subunit and its effects on endocytosis. Embo J 1999; 18:2489-2499.
53. Levkowitz G, Waterman H, Ettenberg SA et al. Ubiquitin ligase activity and tyrosine phosphorylation underlie suppression of growth factor signaling by c-Cbl/Sli-1. Mol Cell 1999; 4:1029-1040.
54. Waterman H, Levkowitz G, Alroy I et al. The RING finger of c-Cbl mediates desensitization of the epidermal growth factor receptor. J Biol Chem 1999; 274:22151-22154.
55. Yokouchi M, Kondo T, Houghton A et al. Ligand-induced ubiquitination of the epidermal growth factor receptor involves the interaction of the c-Cbl RING finger and UbcH7. J Biol Chem 1999; 274:31707-31712.
56. Yoon CH, Lee J, Jongeward GD et al. Similarity of sli-1, a regulator of vulval development in C. elegans, to the mammalian proto-oncogene c-Cbl. Science 1995; 269:1102-1105.
57. de Melker AA, van der Horst G, Calafat J et al. c-Cbl ubiquitinates the EGF receptor at the plasma membrane and remains receptor associated throughout the endocytic route. J Cell Sci 2001; 114:2167-2178.
58. Jiang X, Sorkin A. Epidermal growth factor receptor internalization through clathrin-coated pits requires Cbl RING finger and proline-rich domains but not receptor p. Traffic 2003; 4:529-543.
59. Jiang X, Huang F, Marusyk A et al. Grb2 Regulates internalization of EGF receptors through clathrin-coated pits. Mol Biol Cell 2003; 14:858-870.
60. Levkowitz G, Waterman H, Zamir E et al. c-Cbl/Sli-1 regulates endocytic sorting and ubiquitination of the epidermal growth factor receptor. Genes Dev 1998; 12:3663-3674.
61. Stang E, Johannessen LE, Knardal SL et al. Polyubiquitination of the epidermal growth factor receptor occurs at the plasma membrane upon ligand-induced activation. J Biol Chem 2000; 275:13940-13947.

62. Thien CB, Walker F, Langdon WY. RING finger mutations that abolish c-Cbl-directed polyubiquitination and downregulation of the EGF receptor are insufficient for cell transformation. Mol Cell 2001; 7:355-365.
63. Waterman H, Katz M, Rubin C et al. A mutant EGF-receptor defective in ubiquitylation and endocytosis unveils a role for Grb2 in negative signaling. Embo J 2002; 21:303-313.
64. Duan L, Miura Y, Dimri M et al. Cbl-mediated ubiquitinylation is required for lysosomal sorting of EGF receptor but is dispensable for endocytosis. J Biol Chem 2003; 278:28950-28960.
65. Wang Z, Moran MF. Requirement for the adapter protein GRB2 in EGF receptor endocytosis. Science 1996; 272:1935-1938.
66. Petrelli A, Gilestro GF, Lanzardo S et al. The endophilin-CIN85-Cbl complex mediates ligand-dependent downregulation of c-Met. Nature 2002; 416:187-190.
67. Soubeyran P, Kowanetz K, Szymkiewicz I et al. Cbl-CIN85-endophilin complex mediates ligand-induced downregulation of EGF receptors. Nature 2002; 416:183-187.
68. Springael JY, De Craene JO, Andre B. The yeast Npi1/Rsp5 ubiquitin ligase lacking its N-terminal C2 domain is competent for ubiquitination but not for subsequent endocytosis of the gap1 permease. Biochem Biophys Res Commun 1999; 257:561-566.
69. Katz M, Shtiegman K, Tal-Or P et al. Ligand-independent degradation of epidermal growth factor receptor involves receptor ubiquitylation and Hgs, an adaptor whose ubiquitin-interacting motif targets ubiquitylation by Nedd4. Traffic 2002; 3:740-751.
70. Klapisz E, Sorokina I, Lemeer S et al. A ubiquitin-interacting motif (UIM) is essential for Eps15 and Eps15R ubiquitination. J Biol Chem 2002; 277:30746-30753.
71. Oldham CE, Mohney RP, Miller SL et al. The ubiquitin-interacting motifs target the endocytic adaptor protein epsin for ubiquitination. Curr Biol 2002; 12:1112-1116.
72. Polo S, Sigismund S, Faretta M et al. A single motif responsible for ubiquitin recognition and monoubiquitination in endocytic proteins. Nature 2002; 416:451-455.
73. Raiborg C, Bache KG, Gillooly DJ et al. Hrs sorts ubiquitinated proteins into clathrin-coated microdomains of early endosomes. Nat Cell Biol 2002; 4:394-398.
74. Shih SC, Katzmann DJ, Schnell JD et al. Epsins and Vps27p/Hrs contain ubiquitin-binding domains that function in receptor endocytosis. Nat Cell Biol 2002; 4:389-393.
75. Torrisi MR, Lotti LV, Belleudi F et al. Eps15 is recruited to the plasma membrane upon epidermal growth factor receptor activation and localizes to components of the endocytic pathway during receptor internalization. Mol Biol Cell 1999; 10:417-434.
76. van Delft S, Govers R, Strous GJ et al. Epidermal growth factor induces ubiquitination of Eps15. J Biol Chem 1997; 272:14013-14016.
77. Di Fiore PP, Polo S, Hofmann K. When ubiquitin meets ubiquitin receptors: A signalling connection. Nat Rev Mol Cell Biol 2003; 4:491-497.
78. Shekhtman A, Cowburn D. A ubiquitin-interacting motif from Hrs binds to and occludes the ubiquitin surface necessary for polyubiquitination in monoubiquitinated proteins. Biochem Biophys Res Commun 2002; 296:1222-1227.
79. Gruenberg J, Maxfield FR. Membrane transport in the endocytic pathway. Curr Opinion Cell Biol 1995; 7:552-563.
80. Dupre S, Volland C, Haguenauer-Tsapis R. Membrane transport: Ubiquitylation in endosomal sorting. Curr Biol 2001; 11:R932-934.
81. Komada M, Kitamura N. Hrs and hbp: Possible regulators of endocytosis and exocytosis. Biochem Biophys Res Commun 2001; 281:1065-1069.
82. Zerial M, McBride H. Rab proteins as membrane organizers. Nat Rev Mol Cell Biol 2001; 2:107-117.
83. Barbieri MA, Fernandez-Pol S, Hunker C et al. Role of rab5 in EGF receptor-mediated signal transduction. Eur J Cell Biol 2004; 83:305-314.
84. Benmerah A. Endocytosis: Signaling from endocytic membranes to the nucleus. Curr Biol 2004; 14:R314-316.
85. Miaczynska M, Christoforidis S, Giner A et al. APPL proteins link Rab5 to nuclear signal transduction via an endosomal compartment. Cell 2004; 116:445-456.
86. Stenmark H, Aasland R. FYVE-finger proteins-effector of an inositol lipid. J Cell Science 1999; 112:4175-4183.
87. Carlton J, Bujny M, Peter BJ et al. Sorting nexin-1 mediates tubular endosome-to-TGN transport through coincidence sensing of high- Curvature membranes and 3-phosphoinositides. Curr Biol 2004; 14:1791-1800.
88. Choi JH, Hong WP, Kim MJ et al. Sorting nexin 16 regulates EGF receptor trafficking by phosphatidylinositol-3-phosphate interaction with the Phox domain. J Cell Sci 2004; 117:4209-4218.

89. Asao H, Sasaki Y, Arita T et al. Hrs is associated with STAM, a signal-transducing adaptor molecule. Its suppressive effect on cytokine-induced cell growth. J Biol Chem 1997; 272:32785-32791.
90. Takata H, Kato M, Denda K et al. A hrs binding protein having a Src homology 3 domain is involved in intracellular degradation of growth factors and their receptors. Genes Cells 2000; 5:57-69.
91. Mizuno E, Kawahata K, Kato M et al. STAM proteins bind ubiquitinated proteins on the early endosome via the VHS domain and ubiquitin-interacting motif. Mol Biol Cell 2003; 14:3675-3689.
92. Bache KG, Raiborg C, Mehlum A et al. STAM and Hrs are subunits of a multivalent ubiquitin-binding complex on early endosomes. J Biol Chem 2003; 278:12513-12521.
93. Seykora JT, Mei L, Dotto GP et al. 'Srcasm: A novel Src activating and signaling molecule. J Biol Chem 2002; 277:2812-2822.
94. Katoh Y, Shiba Y, Mitsuhashi H et al. Tollip and Tom1 form a complex and recruit ubiquitin-conjugated proteins onto early endosomes. J Biol Chem 2004; 279:24435-24443.
95. Katzmann DJ, Odorizzi G, Emr SD. Receptor downregulation and multivesicular-body sorting. Nat Rev Mol Cell Biol 2002; 3:893-905.
96. Katzmann DJ, Babst M, Emr SD. Ubiquitin-dependent sorting into the multivesicular body pathway requires the function of a conserved endosomal protein sorting complex, ESCRT-I. Cell 2001; 106:145-155.
97. Li L, Cohen SN. Tsg101: A novel tumor susceptibility gene isolated by controlled homozygous functional knockout of allelic loci in mammalian cells. Cell 1996; 85:319-329.
98. Pornillos O, Alam SL, Rich RL et al. Structure and functional interactions of the Tsg101 UEV domain. Embo J 2002; 21:2397-2406.
99. Babst M, Odorizzi G, Estepa EJ et al. Mammalian tumor susceptibility gene 101 (TSG101) and the yeast homologue, Vps23p, both function in late endosomal trafficking. Traffic 2000; 1:248-258.
100. Lu Q, Hope LW, Brasch M et al. TSG101 interaction with HRS mediates endosomal trafficking and receptor down-regulation. Proc Natl Acad Sci USA 2003; 100:7626-7631.
101. Bache KG, Brech A, Mehlum A et al. Hrs regulates multivesicular body formation via ESCRT recruitment to endosomes. J Cell Biol 2003; 162:435-442.
102. Amit I, Yakir L, Katz M et al. Tal, a Tsg101-specific E3 ubiquitin ligase, regulates receptor endocytosis and retrovirus budding. Genes Dev 2004; 18:1737-1752.
103. Meyer HH, Wang Y, Warren G. Direct binding of ubiquitin conjugates by the mammalian p97 adaptor complexes, p47 and Ufd1-Npl4. EMBO J 2002; 21:5645-5652.
104. Babst M, Katzmann DJ, Snyder WB et al. Endosome-associated complex, ESCRT-II, recruits transport machinery for protein sorting at the multivesicular body. Dev Cell 2002; 3:283-289.
105. Amerik AY, Nowak J, Swaminathan S et al. The Doa4 deubiquitinating enzyme is functionally linked to the vacuolar protein-sorting and endocytic pathways. Mol Biol Cell 2000; 11:3365-3380.
106. Puertollano R. Interactions of TOM1L1 with the multivesicular body sorting machinery. J Biol Chem 2004, (Epub ahead of publication).
107. Jongeward GD, Clandinin TR, Sternberg PW. sli-1, a negative regulator of let-23-mediated signaling in C. elegans. Genetics 1995; 139:1553-1566.
108. French AR, Sudlow GP, Wiley HS et al. Postendocytic trafficking of epidermal growth factor-receptor complexes is mediated through saturable and specific endosomal interactions. J Biol Chem 1994; 269:15749-15755.
109. Bao J, Alroy I, Waterman H et al. Threonine phosphorylation diverts internalized epidermal growth factor receptors from a degradative pathway to the recycling endosome. J Biol Chem 2000; 275:26178-26186.
110. Gáborik Z, Hunyady L. Intracellular trafficking of hormone receptors. Trends Endocrinol Metab 2004; 15:286-293.
111. Cullis DN, Philip B, Baleja JD et al. Rab11-FIP2, an adaptor protein connecting cellular components involved in internalization and recycling of epidermal growth factor receptors. J Biol Chem 2000; 277:49158-49166.
112. Wiley HS. Trafficking of the ErbB receptors and its influence on signaling. Exp Cell Res 2003; 284:78-88.
113. Haugh JM, Huang AC, Wiley HS et al. Internalized epidermal growth factor receptors participate in the activation of p21(ras) in fibroblasts. J Biol Chem 1999; 274:34350-34360.
114. Wang Y, Pennock S, Chen X et al. Endosomal signaling of epidermal growth factor receptor stimulates signal transduction pathways leading to cell survival. Mol Cell Biol 2002; 22:7279-7290.
115. Maier LA, Xu FJ, Hester S et al. Requirements for the internalization of a murine monoclonal antibody directed against the HER-2/neu gene product c-erbB-2. Cancer Res 1991; 51:5361-5369.
116. Mendelsohn J, Baselga J. The EGF receptor family as targets for cancer therapy. Oncogene 2000; 19:6550-6565.
117. Klapper LN, Waterman H, Sela M et al. Tumor-inhibitory antibodies to HER-2/ErbB-2 may act by recruiting c-Cbl and enhancing ubiquitination of HER-2. Cancer Res 2000; 60:3384-3388.

118. Friedman LM, Rinon A, Schechter B et al. Synergistic down-regulation of receptor tyrosine kinases by combinations of mAbs: Implications for cancer immunotherapy. Proc Natl Acad Sci USA 2005; 102:1915-20.

119. Maudsley S, Pierce KL, Zamah AM et al. The beta(2)-adrenergic receptor mediates extracellular signal-regulated kinase activation via assembly of a multi-receptor complex with the epidermal growth factor receptor. J Biol Chem 2000; 275:9572-9580.

120. Oksvold MP, Thien CB, Widerberg J et al. UV-radiation-induced internalization of the epidermal growth factor receptor requires distinct serine and tyrosine residues in the cytoplasmic carboxy-terminal domain. Radiat Res 2004; 161:685-691.

121. Oksvold MP, Skarpen E, Wierod L et al. Relocalization of activated EGF receptor and its signal transducers to multivesicular compartments downstream of early endosomes in response to EGF. Eur J Cell Biol 2001; 80:285-294.

122. He YY, Huang JL, Gentry JB et al. Epidermal growth factor receptor down-regulation induced by UVA in human keratinocytes does not require the receptor kinase activity. J Biol Chem 2003; 278:42457-42465.

123. Ravid T, Sweeney C, Gee P et al. Epidermal growth factor receptor activation under oxidative stress fails to promote c-Cbl mediated down-regulation. J Biol Chem 2002; 277:31214-31219.

CHAPTER 10

Endocytic Trafficking and Human Diseases

Rosa Puertollano*

Abstract

In the last several years an increasing number of genes associated with different human diseases has been identified. Interestingly, many of these genes have been demonstrated to encode components of the intracellular sorting machinery that mediates the selective trafficking of lipids and proteins in the secretory and endocytic pathways. This chapter highlights the molecular basis for selected diseases associated with defects in intracellular trafficking with a specific focus on disorders resulting from aberrant endosomal sorting.

Introduction

The endocytic pathway receives cargo from the cell surface via endocytosis, biosynthetic cargo from the late Golgi complex, and various molecules from the cytoplasm via autophagy. Recently, the intracellular trafficking machinery implicated in delivery of newly synthesized lysosomal proteins from the trans-Golgi network (TGN) to endosomes, fusion of lysosomes with late endosomes, and sorting of membrane proteins into luminal vesicles at endosomes have become topics at the forefront of the study of endosomal biology. The importance of these sorting events is reflected in the fact that defects in the endosomal transport machinery are implicated in a range of human diseases. In this chapter examples of the connection between alterations of the normal trafficking of lipids and proteins along the endosomal pathway and different human pathologies will be described. As criteria for inclusion in this discussion, the protein responsible for the disorder must be a component of the endosomal sorting machinery and the observed cellular defect must arise as a consequence of disturbances in a specific endosomal pathway. The diseases reviewed have been grouped according to the demonstrated or proposed function of the defective protein in intracellular transport (Table 1).

Defects in Protein-Sorting Machinery

One of the best characterized sorting events is the delivery of lysosomal hydrolases from the TGN to endosomes (Fig. 1). After synthesis at the endoplasmic reticulum (ER), lysosomal hydrolases are transported to the Golgi complex where they are posttranslationally modified by the addition of mannose 6-phosphate groups.[1] These groups are then recognized by specific transmembrane receptors named mannose 6-phosphate receptors (MPRs). MPRs contain sorting motifs in their cytosolic tail that allow them to interact with clathrin adaptors[2,3] and be recruited to clathrin-coated areas of the TGN, from which carrier vesicles deliver the MPR-hydrolase complexes to endosomes. The acidic pH within endosomes induces release of the hydrolases from the MPRs, allowing the receptors to return to the TGN for additional rounds of sorting. Fusion events between lysosomes and endosomes could

*Rosa Puertollano—Laboratory of Cell Signaling, National Heart, Lung, and Blood Institute, National Institutes of Health, Bethesda, Maryland 20892, U.S.A. Email: puertolr@mail.nih.gov

Endosomes, edited by Ivan Dikic. ©2006 Landes Bioscience and Springer Science+Business Media.

Table 1. Disorders associated with defects in endosomal trafficking

Disorder	Defective Protein	Protein Function
Defects in protein sorting machinery		
Mucolidosis II or I-cell disease	N-acetylglucosamine 1-phosphotransferase	Addition of mannose-6-phosphate groups to lysosomal enzymes
Hermansky-Pudlak syndrome	β3A subunit of AP3 or HPS	Lysosomal trafficking
Defects in lysosomal biogenesis		
Mucolipidosis IV	h-mucolipin-1	Ca^{2+} channel
Chediak-Higashi syndrome	CHS1	Lysosome fusion/fission
Danon disease	Lamp2	Regulation of autophagy and lysosomal stability
Defects in endosomal Rabs		
Griscelli syndrome	Rab27a	movement of melanosomes and cytotoxic T lymphocyte granule release
Charcot-Marie-Tooth type 2 Neuropathy	Rab7	late endocytic transport and lysosome biogenesis
Choroideremia	Rab escort protein (Rep1)	Geranyl transferase required for Post-translational processing of Rabs proteins
X-linked mental retardation	RabGDP-dissociation inhibitor (GDI)-α	Regulator of Rabs activation
Tuberous sclerosis	Tuberin	Rab5 GTPase activating protein
Defects in lipid trafficking		
Niemann-Pick type C	NPC1 and NPC2	Transport of cholesteryl ester from late endosomes to other organelles
Oculocerebrorenal syndrome of Lowe	inositol polyphosphate 5-phosphatase OCRL-1	conversion of phosphatidyl inositol (4, 5)biphosphate to phosphatidyl inositol 4-phosphate
Autoimmune diseases		
Antiphospholipid syndrome	Autoimmune disease against Lysobisphosphatidic acid	Formation of multivesicular bodies
Stiff-Man syndrome	Autoimmune disease against Amphiphysin I	Regulator of clathrin mediated endocytosis

mediate the final delivery of hydrolases to the lumen of lysosomes where they participate in the degradation of different substrates.

Mucolipidosis Type II or I-Cell Disease

Mucolipidosis type II or I-cell disease is an allelic disorder caused by a deficiency in uridine diphospho (UDP)-*N*-acetylglucosamine: *N*-acetylglucosaminyl-1-phosphotransferase, one of the enzymes that participate in the addition of mannose 6-phosphate to lysosomal hydrolases within the Golgi.[4] Symptoms associated with this disorder include multiple skeletal abnormalities (e.g., congenital hip dislocation and dwarfism), hepatosplenomegaly, and mental retardation.[5,6] The symptoms become obvious during infancy and may include multiple

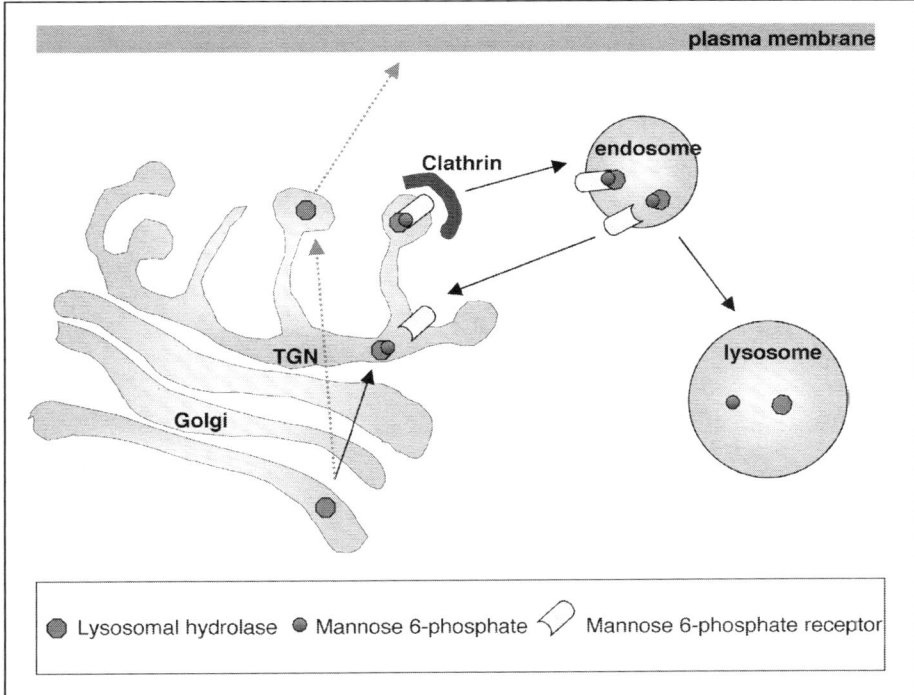

Figure 1. Sorting of lysosomal hydrolases. Addition of mannose 6 phosphate residues allows newly synthesized lysosomal hydrolases to interact with specific receptors called MPRs and be delivered to endosomes through a clathrin dependent pathway (black arrows). Once at endosomes hydrolases dissociate from MPRs and travel to lysosomes while receptors come back to the TGN and participate in additional rounds of transport. In patients with I-cell disease, the inability of hydrolases to receive mannose 6 phosphate residues prevents their interaction with MPRs. Enzymes cannot be transported to lysosomes and are instead secreted into the medium (hatched arrows).

abnormalities of the skull and face and growth delays. Early enzymologic studies showed that cultured fibroblasts from patients were deficient in a number of lysosomal enzymes. Furthermore, these enzymes were found to be present in excess in tissue culture media and in extracellular fluids, such as serum and urine.[7,8] In addition, patient fibroblasts present characteristic phase-dense intracytoplasmic inclusions originated by accumulation of undegraded products. These fibroblasts were named inclusion-cells or I-cells giving name to the disease. Importantly, some cell types, such as hepatocytes and Kupffer cells, seem to have a normal content of lysosomal enzymes despite the absence of phosphotransferase activity[9] indicating that alternative pathways for the trafficking of hydrolases may exist.

Hermansky-Pudlak Syndrome

Hermansky-Pudlak syndrome (HPS) includes a group of several autosomal recessive disorders characterized by oculocutaneous albinism, clotting defects, and storage of ceroid-like material resulting from abnormal function of lysosomes, platelet-dense granules, and melanosomes.[10,11] To date, mutations in seven human genes (HPS1-7) are known to induce HPS. Recent studies have revealed that most of these genes encode proteins that participate in the formation of three different complexes termed lysosome-related organelles complex (BLOC)-1, -2, and -3.[12] It is thought that BLOCs regulate lysosomes and lysosomes-related organelles biogenesis, though they may also be involved in the movement and distribution of late endosomes and lysosomes within the cell.[13] In contrast, HPS-2 patients contain mutations in the β3A subunit of the heterotetrameric complex AP-3 (adaptor protein-3). Four AP complexes have been identified (AP-1, AP-2, AP-3, and AP-4) that regulate sorting at different cellular compartments through mediating the formation of coated transport vesicles, as well as the selection of vesicle cargo.[14] AP-3 is involved in the trafficking of lysosomal proteins from the TGN (and probably also early endosomes) to lysosomes, as evidenced by the fact that several lysosomal membrane proteins (lamp-1, lamp-2, CD63, and limp-II) are missorted to the cell surface in AP-3-deficient mammalian cells.[15,16] In addition, mice deficient in a β3B, an isoform that is exclusively expressed in brain, suffer from spontaneous epileptic seizures and display morphological abnormalities at synapses, suggesting that AP-3B might regulates the formation and function of a subset of synaptic vesicles in the brain.[17]

Defects in Lysosomal Biogenesis

In the last several years, numerous models have been proposed to explain how material targeted for lysosomal degradation is delivered from late endosomes to lysosomes.[18] For example, the endosomal maturation theory was updated to the kiss-and-run premise, in which multiple restricted fusion events would take place between late endosomes and lysosomes allowing delivery of luminal components while maintaining the integrity of the limiting membranes.[19] Alternatively, it has been proposed that a complete fusion between these two compartments might also occur, resulting in the formation of endosome-lysosome hybrid organelles. Lysosomes would be reformed from the hybrid in a process that requires recycling of membrane proteins to endosomes or the TGN and condensation of the intraluminal content. Interestingly, such hybrid organelles have been observed both in vivo and in vitro[20,21] and may in fact act as the major site for hydrolysis of endocytosed molecules. The machinery implicated in the fusion between late endosomes and lysosomes has not been characterized, but indirect evidences suggest that it could involve the small GTPase Rab 7;[22] a tethering complex formed by the mammalian homologs of the yeast proteins Vps11p, Vps16p, Vps18p, Vps33p, Vps39p, and Vps41p;[23,24] and specific SNAREs.[25] Calcium also seems to play a fundamental role for fusion and lysosomal reformation.[26,27]

Mucolipidosis Type IV

One of the disorders associated to defects in the late steps of endocytosis and lysosomal biogenesis is Mucolipidosis type IV (MLIV). MLIV is an autosomal recessive lysosome storage disorder characterized by severe psychomotor retardation and oftalmological abnormalities, including corneal opacity, retinal degeneration, and strabismus.[28] Patient's cells often contain enlarged vacuolar structures that accumulate sphingolipids, phospholipids and mucopolysacharides and display a higher pH than normal lysosomes. Interestingly, MCOLN1, the gene mutated in MLIV patients, encodes a protein termed h-mucolipin-1[29-31] that functions as a Ca^{2+} permeable channel and can be modulated by changes in pH and Ca^{2+} concentration.[32] It has been proposed that h-mucolipin-1 may be involved in the regulation of lysosomes biogenesis, and more specifically in the reformation of lysosomes from hybrid organelles. In agreement with this idea, recent experiments have shown that mutants of cup-5, the *C. elegans* orthologue of h-mucolipin-1, are associated with accumulation of endocytosed green fluorescent protein in enlarged vacuoles. Importantly these structures resemble endosome–lysosome hybrid organelles, based on the presence of RME-8 and LMP-1 (two distinct markers of late endosomes and lysosomes, respectively).[33,34]

Chediak-Higashi Syndrome

Chediak-Higashi syndrome is a disorder characterized by defects in blood clotting and pigmentation. However, in contrast with HPS, Chediak-Higashi patients also show neurologic dysfunctions and immunological deficits, including accumulation of large lysosomal granules in leukocytes, large eosinophilic, neutropenia, increased susceptibility to infection, and abnormal malignant lymphoma.[35] As a result, death often occurs before the age of seven years. The protein defective in this disorder is CHS1[36] and, although its function is still unknown, it has been reported that over expression of different CHS1 domains in Cos-7 cells causes dramatic changes in the size of lysosomes.[37] Since the fusion of secretory lysosomes with the plasma membrane is also inhibited in CHS1 defective cells, it was proposed that this protein probably regulates membrane fusion/fission events. This is in agreement with a recent study that describes an interaction between CHS1 and a soluble SNARE complex protein implicated in membrane fusion.[38]

Danon Disease

It has also been suggested that some lysosomal membrane glycoproteins (LPGs, that include lysosome-associated membrane proteins or LAMPs and lysosomal integral membrane proteins or LIMPs) could be implicated in lysosomal biogenesis.[39] These proteins are major components of the limiting membrane of lysosomes, but are also present in late endosomes. Based on their heavy glycosylation it was assumed that LPGs main function is to protect membranes from degradation by lysosomal hydrolases. LAMP-2 deficiency leads to Danon disease in humans, a pathology characterized by cardiomyopathy, myophaty and variable mental retardation.[40] Electron microscopy studies reveled a massive accumulation of authophagic vacuoles in numerous tissues, including hearth, muscle, and liver suggesting that LAMP-2 might play a role in autophagy and lysosomal stability. In addition, it has been reported that LIMP-2/LGP85 over expression causes accumulation of enlarged hybrid organelles while depletion of this protein in mice results in deafness, urogenital track obstruction, and peripheral neuropathy.[41]

Defects in Function of Endosomal Rabs

Rab proteins are small monomeric GTPases with molecular masses in the 20–30 kDa range. Multiple Rabs have been shown to participate in the formation, fusion and movement of vesicular traffic intermediaries between different membrane compartments of the cell.[42-45] Rabs function as molecular switches by cycling between two interconvertible forms, a cytosolic GDP-bound (inactive) form and a membrane associated GTP-bound (active) form. In addition, Rab proteins contain unique, hypervariable C-terminal domains with either one or, more frequently, two Cys residues both of which are modified by geranylgeranyl groups. This prenylation occurs in several steps and is essential for Rab association with intracellular membranes. Newly synthesized Rabs bind to a 95 kDa protein named Rab escort protein (REP) forming a stable complex that is then recognized by a Rab geranylgeranyl transferase (RabGGT), which catalyses the transference of geranylgeranyl groups to the Rabs C-terminal cysteines. Prenylated Rabs can be delivered to their appropriate donor membrane where specific Rab guanine nucleotide exchange factors (GEF) replace GDP with GTP (Fig. 2, steps 1-5). Active GTP-bound Rabs can now recruit different effectors and exert their specific functions that may include cargo selection and budding, as well as movement, tethering and fusion of vesicles at their final destinations. Finally, Rab GTPase activating proteins (GAP) promote GTP hydrolysis and render Rabs into an inactive GDP-bound conformation. Removal of inactive Rabs from the target membranes is mediated by Rab GDP-dissociation inhibitor (GDI) that can maintain Rabs in a GDP-bound inactive form in the cytosol or recycle them back to donor membranes (Fig. 2, steps 7-9).

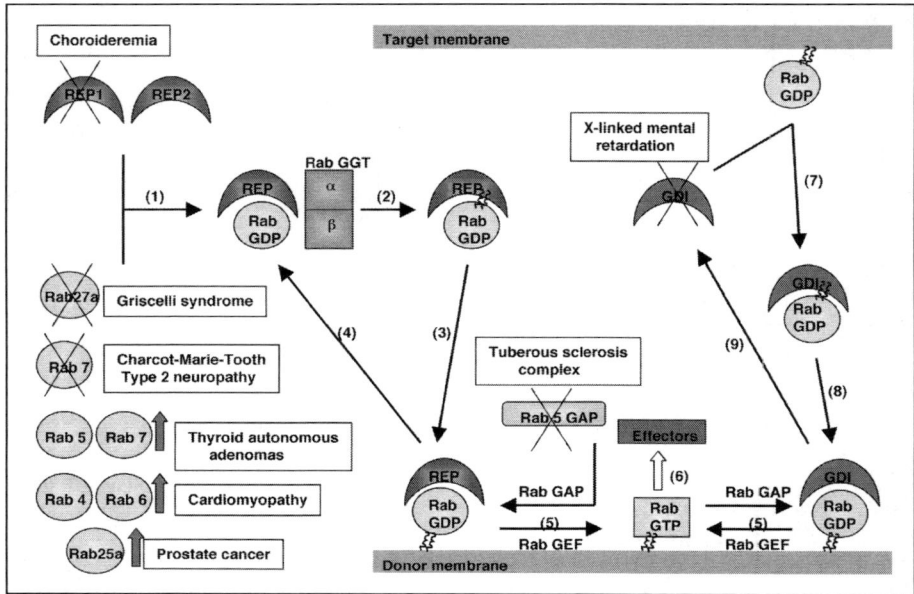

Figure 2. Diseases associated with defects in endosomal Rabs. After synthesis in the cytosol, Rab proteins form a stable complex with REP (1). RabGGT can recognize this complex and catalyze the transference of geranylgeranyl groups to the C-terminal region of Rabs (2). Prenylated Rabs are delivered to donor membranes (3) where specific GEFs promote the exchange of GDP by GTP (5) activating the Rabs that can now recruit effectors and exert their functions (6). In addition, GAP proteins stimulate GTP hydrolysis and render Rabs in a GDP-bound inactive state (5). RabGDI retrieves inactive Rab proteins from the target membrane and maintains them in an inactive state in the cytosol (7) or delivers them back to the donor membrane (8). Several human diseases caused by defects (cross) or over expression (gray arrows) of different Rabs and Rabs regulators are indicated.

Importantly, numerous pathologies have been associated with loss of function of Rabs and Rab regulators, as well as Rab overexpression. These are described below and in Figure 2.

Pathologies Associated with Loss of Function of Rabs and Rabs Regulators

Griscelli Syndrome

Griscelli syndrome is rare autosomal recessive disease induced by the absence of Rab27a.[46-48] In melanocytes Rab27a promotes the recruitment of myosin Va and melanophilin to the melanosome membrane, thus allowing interaction of melanosomes with the actin cytoskeleton. This interaction mediates the concentration of melanosomes in the tip of melanocyes dendrites and the transference of pigments to keratinocytes.[49-50] Absence of Rab27a causes a dramatic alteration of melanosomes trafficking, and accumulation of melanosomes in the central region of melanocytes, resulting in defects in hair and skin pigmentation and partial albinism.[51] In addition, defects in Rab27a expression also correlate with impaired immunological response due to defects in the secretion of lytic granules by cytotoxic T lymphocytes (CTLs).[52,53] In some patients neurological alterations has also been reported, although this is thought to be a secondary effect consequence of uncontrolled activation of T-lymphocytes and macrophages and subsequent leukocyte infiltration of the brain.[54]

Charcot-Marie-Tooth Type 2 Neuropathy

Mutations that disrupt Rab7 function are responsible for Charcot-Marie-Tooth type 2 neuropathy.[55] This disease is characterized by a severe motor and sensory neurons impairment, distal muscle weakness and high frequency of foot ulcers that often require amputation because of recurrent infections. Interestingly, mutations in myotubularin-related protein 2, a phosphatase implicated in the metabolism of phosphatidylinositol (3,5) biphosphate and phosphatidylinositol 3-phosphate;[56] and mutations in KIF1B, a kinase that regulates synaptic transport,[57] also result in different forms of Charcot-Marie-Tooth disease, indicating that Rab7, myotubularin, and KIF1B may participate in the same pathway.

Choroideremia

Choroideremia (CHM) is a X-linked disease characterized by a slow degeneration of photorepectors, choriocapillaris, and retinal pigment ephitelium. This pathology usually starts in the second or third decade of life by progressive night blindness and loss of peripheral vision and often results in complete blindness.[58,59] CHM is caused by loss-of-function mutations in REP-1, a protein that regulates Rab prenylation (see above). However, it is thought that this mutation may be partially compensated by REP-2. This could explain why CHM phenotype is restricted to some very specific cells types and predicts the existence of retina-specific factors that interact with REP-1.

In mice reduction in the activity of RabGGT activity correlates with clotting disorders, including prolonged bleeding times, thrombocytopenia, and reduced platelet granule content, as well as with albinism and inability of CTLs to polarize lytic granules toward the immunological synapse with the target.

X-Linked Mental Retardation

Mutations in RabGDIα have been found in patients with X-linked mental retardation.[60] Mammalian Rab GDI consists of three members: Rab GDI α, β, and γ. Rab GDIα is specifically expressed in neuronal tissue where regulates the activity of Rabs implicated in neurotransmission, such as Rab3a.[61] Rab GDIα-deficient mice display neuronal hypersensitivity and high susceptibility to suffer epileptic seizures,[62] indicating that Rab GDIα might play an important role as a negative regulator of the synaptic function.

Tuberous Sclerosis Complex

Tuberous sclerosis complex (TSC) is an autosomal dominant disorder characterized by the presence of benign tumors called hamartomas in multiple organs, including the central nervous system, lung, kidney, heart, and skin; learning and behavioral difficulties; and renal complications.[63] Mutation of two genes, TSC1 and TSC2, result in this clinical disorder. Importantly, TSC2 gene encodes tuberin, a GAP protein that stimulates GTP hydrolysis on Rab5,[64] suggesting that aberrations in the Rab5 dependent endocytic pathway might be linked to this disease.

Pathologies Associated to Rabs Overexpression

Alterations in the level of expression of a variety of Rabs have also been related to several human diseases. In some cases the overexpression may be consequence of somatic rearrangements while in other cases they are a consequence of sustained intracellular signaling. High levels of Rab5 and Rab7 have been associated with a certain type of benign thyroid tumors called thyroid autonomous adenomas (AA). It appears that this overexpression promotes an increased association of Rab5 and Rab7 with endosomes, resulting in augmented processing of thyroglobulin and thyroid hormone production.[65] Rab 7 over expression has also been observed in a mouse model for atherogenesis;[66] while the level of Rab 25, a Rab implicated in recycling to the apical membrane, is altered in prostate cancer cell lines.[67] It has also been noted that the levels of Rab6, which is involved in microtubule dependent pathways through the

Golgi apparatus and from endosomes to Golgi, as well as the levels of the endosomal Rab4 are upregulated in a dilated cardiomyopathy model overexpressing β2-adrenergic receptors.[68] Finally, increased expression of a Rab5 GAP named PRC17 has been reported in metastatic prostate tumors, indicating that alterations in the levels of Rab regulators may also account for a number of human diseases.[69]

Defects in Lipid Trafficking

Niemann-Pick Type C Disease

Cholesterol is a major structural component of mammalian cellular membranes and also acts as a regulator of lipogenic gene expression and membrane protein function. Cells have to carefully regulate the amount and distribution of cholesterol to assure proper function. The principal disorder associated with defects in cholesterol trafficking at endosomes is Niemann-Pick type C (NPC) disease.[70] This syndrome is characterized by progressive neurological degeneration and is often associated to hepatosplenomegaly. Patients usually appear normal for the first two years of life with symptoms appearing between 2 and 4 years. At the cellular level there is a clear accumulation of abnormal amounts of cholesterol and other lipids (including glycosphingolipids (GSLs), sphingomyelin, lysobisphosphatidic acid (LBPA), and phospholipids) in a late endosomal/lysosomal compartment. It is interesting to note that in neurons the total amount of cholesterol is not significantly different between NPC patients and controls, however, its distribution is dramatically altered. Cholesterol accumulates in enlarged endosomal/lysosomal structures located in the neurons cell body while it is absent from endosomal organelles at the distal axon.

95% of NPC patients contain mutations in NPC1, a multi-spanning transmembrane protein that localizes to late endosomes/lysosomes and is thought to regulate the trafficking of free cholesterol from late endosomes to other cell compartments, including the plasma membrane and endoplasmic reticulum (ER).[71,72] The remaining 5% of NPC patients is defective on NPC2, a soluble cholesterol binding protein that normally cycles between TGN and endosomes through a CI-MPR dependent pathway, but accumulates in late endosomes in NPC patients.[73] It is thought that accumulation of cholesterol in late endosomes resulting from mutations in either NPC1 or NPC2 could result in a "lipid traffic jam" that would trap other lipids and transmembrane proteins and disturb the cellular cholesterol homeostatic response. Recent experiments have shown that over expression of Rab 7 or Rab 9, but not Rab11, corrects defective lipid trafficking and abrogates cholesterol storage in NPC cells.[74] The mechanism that mediates this restoration if unknown though it has been suggested that Rab 7 and Rab 9 over expression could increase the transport of accumulated lipids from late endosomes.

Recently, some authors have pointed out similarities between NPC and Alzheimer's disease (AD).[75] These similarities include endosome enlargement, elevated hydrolase content, and accumulation of β-cleaved amyloid precursor protein (APP) and Aβ peptides within endosomes, and suggest that defects on endosomal trafficking might lead to defective processing of APP and neurodegeneration.

Oculocerebrorenal Syndrome of Lowe

Another pathology associated with defects in lipid trafficking is the oculocerebrorenal syndrome of Lowe (OCRL), a rare X linked disorder characterized by mental retardation, congenital cataracts, and reduced ammonia production by the kidney.[76] The gene mutated in this syndrome encodes a phosphatidylinositol (4, 5) biphosphate 5-phosphatase named OCRL-1 that catalyzes the conversion of phosphatidylinositol (4, 5) biphosphate to phosphatidylinositol 4-phosphate.[77] Fibroblasts from patients with Lowe syndrome showed abnormalities of the actin cytoskeleton as well as atypical distribution of gensolin and alpha-actinin, two actin-binding proteins regulated by calcium and phosphatidylinositol (4, 5) biphosphate.[78] Actin is necessary for formation and maintenance of tight junctions. Since these structures play a crucial role in the function of renal proximal tubule and lens differentiation it has been proposed that defects in actin polymerization could account for the OCRL phenotype. However, it is important to note that phosphatidylinositol 4-phosphate also participates in the recruitment of clathrin adaptors to Golgi membranes. It has been described that depletion of a major phosphatidylinositol 4-kinase by siRNA causes AP-1 to become citosolic, an effect that can be reverted by adding exogenous phosphatidylinositol 4-phosphate.[79] As mentioned above, clathrin adaptors mediate transport of different proteins, including lysosomal hydrolases, from TGN to endosomes, therefore, defects in OCRL-1 could disrupt the normal endosomal trafficking.

Autoimmune Diseases

Autoantibodies have traditionally been used as diagnostic markers for various autoimmune diseases. Interestingly, some of these autoantibodies recognize endosomal proteins and their presence correlates with specific disorders. For example, autoantibodies to early endosomal antigen (EEA1), a protein located on the cytosolic face of early endosomes and implicated in regulation of endosome fusion,[80,81] have been reported in the sera of patients with neurological disorders.[82,83] Early endosomes are key functional components of both presynaptic and post-synaptic neurons indicating that the presence of autoantibodies against EEA1 could induce aberrations in the endosomal pathway and defects in neurotransmission.

The antiphospholipid syndrome (APS) is a disorder in which antibodies binding to phospholipids (PL) are thought to be involved in the development of thrombosis and/or pregnancy complications.[84] It has been described that the sera of some of these patients contain antibodies against lysobisphosphatidic acid (LBPA), a lipid highly enriched in late endosomes that regulates the formation of multivesicular bodies (MVBs).[85,86] Moreover, binding of specific antibodies to LBPA induces an alteration in the structure and function of late endosomes.[87]

There are other examples of autoimmune diseases associated with the presence of antibodies directed against endosomal proteins. Stiff-Man syndrome, a rare central nervous system disorder characterized by muscular rigidity and episodic spasms[88] correlates in a subset of patients with the presence of anti-amphiphysin autoantibodies;[89] necrotizing and crescentic glomerulonephritis (NCGN) is frequently associated with circulating autoantibodies against Lamp-2;[90] and CLIP-170, a protein that connect endosomes with cytosolic microtubules,[91] has been identified as a new autoantigen in three patients suffering from limited scleroderma, glioblastoma and idiopathic pleural effusion.[92]

Conclusions and Future Prospects

The recent identification of new components of the sorting machinery has allowed a better understanding of the molecular mechanism of different human diseases. At the same time the identification of disorders caused by defects in intracellular trafficking provides very valuable information about the role of specific proteins in a particular pathway. Some of the examples included in this chapter show how defects in different proteins can cause similar phenotypic effects. In addition, the presence of components with redundant functions may mask the defect of a protein or limit the effect of this mutation to specific cell types. Certainly, a better comprehension of the mechanisms that regulate intracellular trafficking will help to design improved strategies to fight numerous human diseases.

Acknowledgements

My thanks to all members of my laboratory for helpful comments, and to C. Mullins for critical reading of the manuscripts and valuable suggestions.

References

1. Kornfeld S. Structure and function of the mannose 6-phosphate/insulinlike growth factor II receptors. Annu Rev Biochem 1992; 61:307-330.
2. Puertollano R, Aguilar RC, Gorshkova I et al. Sorting of mannose 6-phosphate receptors mediated by the GGAs. Science 2001; 292:1712-1716.
3. Doray B, Ghosh P, Griffith J et al. Cooperation of GGAs and AP-1 in packaging MPRs at the trans-Golgi network. Science 2002; 297:1700-1703.
4. Varki A, Reitman ML, Kornfeld S. The enzymatic defect in I-cell disease (ML II) and pseudo-Hurler polydystrophy (ML III). Clin Res 1981; 29:514A.
5. Leroy JG, DeMars RI. Mutant enzymatic and cytological phenotypes in cultured human fibroblasts. Science 1967; 157:804-806.
6. Spranger JW, Weidemann HR. The genetic mucolipidoses. Human genetik 1970; 9:113-139.
7. Tondeur M, Vamos-Hurwitz E, Mockel-Pohl S et al. Clinical, biochemical, and ultrastructural studies in a case of chondrodystrophy presenting the I-cell phenotype in culture. J Pediatr 1971; 79:366-378.
8. Lightbody J, Weissmann U, Hadorn B et al. I-cell disease: Multiple lysosomal-enzyme defect. Lancet 1971; 1:451.
9. Kornfeld S, Sly WS. I-cell disease and pseudo-Hurler polydystrophy: Disorders of lysosomal enzyme phosphorylation and localization. In: Scriver CR, Beaudet AL, Sly WS, Valle D, eds. The metabolic and molecular bases of inherited disease. New York: McGraw-Hill, 1995:2495–2508.
10. Hermansky F, Cieslar P, Matousova O et al. Proceedings: Study of albinism in relation to Hermansky-Pudlak syndrome. Thromb Diath Haemorrh 1975; 34:360.
11. Oh J, Ho L, Ala-Mello S et al. Mutation analysis of patients with Hermansky-Pudlak syndrome: A frameshift hot spot in the HPS gene and apparent locus heterogeneity. Am J Hum Genet 1998; 62:593-598.
12. Dell'Angelica EC. The building BLOC(k)s of lysosomes and related organelles. Curr Opin Cell Biol 2004; 16:458-464.
13. Nazarian R, Falcon-Perez JM, Dell'Angelica EC. Biogenesis of lysosome-related organelles complex 3 (BLOC-3): A complex containing the Hermansky-Pudlak syndrome (HPS) proteins HPS1 and HPS4. Proc Natl Acad Sci USA 2003; 100:8770-8775.
14. Robinson MS, Bonifacino JS. Adaptor-related proteins. Curr Opin Cell Biol 2001; 13:444-453.
15. Dell'Angelica EC, Shotelersuk V, Aguilar RC et al. Altered trafficking of lysosomal proteins in HermanskyPudlak syndrome due to mutations in the beta 3A subunit of the AP-3 adaptor. Mol Cell 1999; 3:11-21.
16. Ihrke G, Kyttala A, Russell MR et al. Differential use of two AP-3-mediated pathways by lysosomal membrane proteins. Traffic 2004; 5:946-962.
17. Nakatsu F, Okada M, Mori F et al. Defective function of GABA-containing synaptic vesicles in mice lacking the AP-3B clathrin adaptor. J Cell Biol 2004; 167:293-302.
18. Luzio JP, Poupon V, Lindsay MR et al. Membrane dynamics and the biogenesis of lysosomes. Mol Membr Biol 2003; 20:141-154.
19. Valtorta F, Meldolesi J, Fesce R. Synaptic vesicles: Is kissing a matter of competence? Trends Cell Biol 2001; 11:324-328.
20. Bright NA, Reaves BJ, Mullock BM et al. Dense core lysosomes can fuse with late endosomes and are reformed from the resultant hybrid organelles. J Cell Sci 1997; 110:2027-2040.
21. Mullock BM, Bright NA, Fearon CW et al. Fusion of lysosomes with late endosomes produces a hybrid organelle of intermediate density and is NSF dependent. J Cell Biol 1998; 140:591-601.
22. Wickner W. Yeast vacuoles and membrane fusion pathways. EMBO J 2002; 21:1241-1247.
23. Seals DF, Eitzen G, Margolis N et al. A Ypt/Rab effector complex containing the Sec I homolog Vps33p is required for homotypic vacuole fusion. Proc Natl Acad Sci USA 2000; 97:9402-9407.
24. Wurmser AE, Sato TK, Emr SD. New component of the vacuolar class C-Vps complex couples nucleotide exchange on the Ypt7 GTPase to SNARE dependent docking and fusion. J Cell Biol 2000; 151:551-562.
25. Antonin W, Holroyd C, Fasshauer D et al. A SNARE complex mediating fusion of late endosomes defines conserved properties of SNARE structure and function. EMBO J 2000; 19:6453-6464.
26. Merz AJ, Wickner WT. Trans-SNARE interactions elicit Ca2+ efflux from the yeast vacuole lumen. J Cell Biol 2004; 164:195-206.
27. Pryor PR, Mullock BM, Bright NA et al. The role of intraorganellar Ca(2+) in late endosome-lysosome heterotypic fusion and in the reformation of lysosomes from hybrid organelles. J Cell Biol 2000; 149:1053-1062.
28. Bach G. Mucolipidosis type IV. Mol Genet Metab 2001; 73:197-203.

29. Bargal R, Avidan N, Ben-Asher E et al. Identification of the gene causing mucolipidosis type IV. Nat Genet 2000; 26:120-123.

30. Bassi MT, Manzoni M, Monti E et al. Cloning of the gene encoding a novel integral membrane protein, mucolipidin- and identification of the two major founder mutations causing mucolipidosis type IV. Am J Hum Genet 2000; 67:1110-1120.

31. Sun M, Goldin E, Stahl S et al. Mucolipidosis type IV is caused by mutations in a gene encoding a novel transient receptor potential channel. Hum Mol Genet 2000; 9:2471-2478.

32. LaPlante JM, Falardeau J, Sun M et al. Identification and characterization of the single channel function of human mucolipin-1 implicated in mucolipidosis type IV, a disorder affecting the lysosomal pathway. FEBS Lett 2002; 532:183-187.

33. Fares H, Grant B. Deciphering endocytosis in Caenorhabditis elegans. Traffic 2002; 3:11-19.

34. Treusch S, Knuth S, Slaugenhaupt SA et al. Caenorhabditis elegans functional orthologue of human protein h-mucolipin-1 is required for lysosome biogenesis. Proc Natl Acad Sci USA 2004; 101:4483-4488.

35. Barak Y, Nir E. Chediak-Higashi syndrome. Am J Pediatr Hematol Oncol 1987; 9:42-55.

36. Jenkins NA, Justice MJ, Gilbert DJ et al. Nidogen/entactin (Nid) maps to the proximal end of mouse chromosome 13 linked to beige (bg) and identifies a new region of homology between mouse and human chromosomes. Genomics 1991; 9:401-403.

37. Ward DM, Shiflett SL, Huynh D et al. Use of expression constructs to dissect the functional domains of the CHS/beige protein: Identification of multiple phenotypes. Traffic 2003; 4:403-415.

38. Tchernev VT, Mansfield TA, Giot L et al. The Chediak-Higashi protein interacts with SNARE complex and signal transduction proteins. Mol Med 2002; 8:56-64.

39. Eskelinen EL, Tanaka Y, Saftig P. At the acidic edge: Emerging functions for lysosomal membrane proteins. Trends Cell Biol 2003; 13:137-145.

40. Nishino I, Fu J, Tanji K et al. Primary LAMP-2 deficiency causes X-linked vacuolar cardiomyopathy and myopathy (Danon disease). Nature 2000; 406:906-910.

41. Kuronita T, Eskelinen EL, Fujita H et al. A role for the lysosomal membrane protein LGP85 in the biogenesis and maintenance of endosomal and lysosomal morphology. J Cell Sci 2002; 115:4117-4131.

42. Zerial M, McBride H. Rab proteins as membrane organizers. Nature Rev Mol Cell Biol 2001; 2:107-117.

43. Pfeffer SR. Rab GTPases: Specifying and deciphering organelle identity and function. Trends Cell Biol 2001; 11:487-491.

44. Stein MP, Dong J, Wandinger-Ness A. Rab proteins and endocytic trafficking: Potential targets for therapeutic intervention. Adv Drug Deliv Rev 2003; 55:1421-1437.

45. Pfeffer S, Aivazian D. Targeting Rab GTPases to distinct membrane compartments. Nat Rev Mol Cell Biol 2004; 5:886-896.

46. Menasche G, Pastural E, Feldmann J et al. Mutations in RAB27A cause Griscelli syndrome associated with haemophagocytic syndrome. Nat Genet 2000; 25:173-176.

47. Sanal O, Ersoy F, Tezcan I et al. Griscelli disease: Genotypephenotype correlation in an array of clinical heterogeneity. J Clin Immunol 2002; 22:237-243.

48. Bahadoran P, Busca R, Chiaverini C et al. Characterization of the molecular defects in Rab27a, caused by RAB27A missense mutations found in patients with Griscelli syndrome. J Biol Chem 2003; 278:11386-11392.

49. Hume AN, Collinson LM, Rapak A et al. Rab27a regulates the peripheral distribution of melanosomes in melanoctes. J Cell Biol 2001; 152:795-808.

50. Wu X, Rao K, Bowers MB et al. Rab27a enables mosin Va-dependent melanosome capture b recruiting the mosin to the organelle. J Cell Sci 2001; 114:1091-1100.

51. Marks MS, Seabra MC. The melanosome: Membrane dynamics in black and white. Nat Rev Mol Cell Biol 2001; 2:738-748.

52. Haddad EK, Wu X, Hammer IIIrd JA et al. Defective granule exocytosis in Rab27a-deficient lymphocytes from Ashen mice. J Cell Biol 2001; 152:835-842.

53. Stinchcombe JC, Barral DC, Mules EH et al. Rab27a is required for regulated secretion in cytotoxic T lymphocytes. J Cell Biol 2001; 152:825-834.

54. Menasche G, Pastural E, Feldmann J et al. Mutations in RAB27A cause Griscelli syndrome associated with haemophagocytic syndrome. Nat Genet 2000; 25:173-176.

55. Verhoeven K, De Jonghe P, Coen K et al. Mutations in the small GTP-ase late endosomal protein RAB7 cause Charcot – Marie – Tooth type 2B neuropathy. Am J Hum Genet 2003; 72:722-727.

56. Schaletzky J, Dove SK, Short B et al. Phosphatidylinositol-5-phosphate activation and conserved substrate specificity of the myotubularin phosphatidylinositol 3-phosphatases. Curr Biol 2003; 13:504-509.

57. Zhao C, Takita J, Tanaka Y et al. Charcot – Marie –Tooth disease type 2A caused by mutation in a microtubule motor KIF1Bbeta. Cell 2001; 105:587-597.
58. Seabra MC. New insights into the pathogenesis of choroideremia: A tale of two REPs. Ophthalmic Genet 1996; 17:43-46.
59. van den Hurk JA, Schwartz M, van Bokhoven H et al. Molecular basis of choroideremia (CHM): Mutations involving the Rab escort protein-1 (REP-1) gene. Hum Mutat 1997; 9:110-117.
60. D'Adamo P, Menegon A, Lo Nigro C et al. Mutations in GDI1 are responsible for X-linked nonspecific mental retardation. Nat Genet 1998; 19:134-139.
61. Sasaki T, Kikuchi A, Araki S et al. Purification and characterization from bovine brain cytosol of a protein that inhibits the dissociation of GDP from the subsequent binding of GTP to smg p25A, a ras p21-like GTP-binding protein. J Biol Chem 1990; 265:2333-2337.
62. Ishizaki H, Miyoshi J, Kamiya H et al. Role of rab GDP dissociation inhibitor alpha in regulating plasticity of hippocampal neurotransmission. Proc Natl Acad Sci USA 2000; 97:11587-11592.
63. Gomez MR. Tuberous Sclerosis. 2nd ed. New York: Raven Press, 1988.
64. Xiao GH, Shoarinejad F, Jin F et al. The tuberous sclerosis 2 gene product, tuberin, functions as a Rab5 GTPase activating protein (GAP) in modulating endocytosis. J Biol Chem 1997; 272:6097–6100.
65. Croizet-Berger K, Daumerie C, Couvreur M et al. The endocytic catalysts, Rab5a and Rab7, are tandem regulators of thyroid hormone production. Proc Natl Aca Sci USA 2002; 99:8277-8282.
66. Kim JY, Jang MK, Lee SS et al. Rab7 gene is up-regulated by cholesterolrich diet in the liver and artery. Biochem Biophys Res Commun 2002; 293:375-382.
67. Calvo A, Xiao N, Kang J et al. Alterations in gene expression profiles during prostate cancer progression: Functional correlations to tumorigenicity and down-regulation of selenoprotein-P in mouse and human tumors. Cancer Res 2002; 62:5325-5335.
68. Wu G, Yussman MG, Barrett TJ et al. Increased myocardial Rab GTPase expression: A consequence and cause of cardiomyopathy. Circ Res 2001; 89:1130-1137.
69. Pei L, Peng Y, Yang Y et al. PRC17, a novel oncogene encoding a Rab GTPase-activating protein, is amplified in prostate cancer. Cancer Res 2002; 62:5420-5424.
70. Patterson MC, Vanier MT, Suzuki K et al. In: Scriver CR, Beaudet AL, Sly WS, Valle D, eds. The Metabolic and Molecular Bases of Inherited Disease. New York: McGraw-Hill, 2001:3611-3633.
71. Ioannou YA. The structure and function of the Niemann-Pick C1 protein. Mol Genet Metab 2000; 71:175-181.
72. Scott C, Ioannou YA. The NPC1 protein: Structure implies function. Biochim Biophys Acta 2004; 1685:8-13.
73. Blom TS, Linder MD, Snow K et al. Defective endocytic trafficking of NPC1 and NPC2 underlying infantile Niemann-Pick type C disease. Hum Mol Genet 2003; 12:257-272.
74. Choudhury A, Dominguez M, Puri V et al. Rab proteins mediate Golgi transport of caveola-internalized glycosphingolipids and correct lipid trafficking in Niemann–Pick C cells. J Clin Invest 2002; 109:1541-1550.
75. Nixon RA. Niemann-Pick Type C disease and Alzheimer's disease: The APP-endosome connection fattens up. Am J Pathol 2004; 164:757-761.
76. Charnas LR, BernardiniI, Rader D et al. Clinical and laboratory findings in the oculocerebrorenal syndrome of Lowe, with special reference to growth and renal function. N Engl J Med 1991; 324:1318-1325.
77. Attree O, Olivos IM, Okabe I et al. The Lowe's oculocerebrorenal syndrome gene encodes a protein highly homologous to inositol polyphosphate-5-phosphatase. Nature 1992; 358:239-242.
78. Suchy SF, Nussbaum RL. The deficiency of PIP2 5-phosphatase in Lowe syndrome affects actin polymerization. Am J Hum Genet 2002; 71:1420-1427.
79. Wang YJ, Wang J, Sun HQ et al. Phosphatidylinositol 4 phosphate regulates targeting of clathrin adaptor AP-1 complexes to the Golgi. Cell 2003; 114:299-310.
80. Simonsen A, Lippe R, Christoforidis S et al. EEA1 links PI (3) K function to Rab5 regulation of endosome fusion. Nature 1998; 394:494-498.
81. Rubino M, Miaczynska M, Lippe R et al. Selective membrane recruitment of EEA1 suggests a role in directional transport of clathrin-coated vesicles to earl endosomes, J Biol Chem 2000; 275:3745-3748.
82. Selak S, Schoenroth L, Senécal J-L at al. Early endosome antigen 1: An autoantigen associated with neurological diseases. J Invest Med 1999; 47:311-318.
83. Selak S, Mahler M, Miyachi K et al. Identification of the B-cell epitopes of the early endosome antigen 1 (EEA1). Clin Immunol 2003; 109:154-164.
84. Arnout J, Vermylen J. Current status and implications of autoimmune antiphospholipid antibodies in relation to thrombotic disease. J Thromb Haemost 2003; 1:931-942.

85. Kobayashi T, Stang E, Fang KS et al. A lipid associated with the antiphospholipid syndrome regulates endosome structure and function. Nature 1998; 392:193-197.
86. Kobayashi T, Beuchat MH, Lindsay M et al. Late endosomal membranes rich in lysobisphosphatidic acid regulate cholesterol transport. Nat Cell Biol 1999; 1:113-118.
87. Schmid SL, Cullis PR. Endosome marker is fat not fiction. Nature 1998; 392:135-136.
88. Brown P, Marsden CD. The stiff man and stiff man plus syndromes. J Neurol 1999; 246:648-652.
89. Petzold GC, Marcucci M, Butler MH et al. Rhabdomyolysis and paraneoplastic stiff-man syndrome with amphiphysin autoimmunity. Ann Neurol 2004; 55:286-290.
90. Kain R, Matsui K, Exner M et al. A novel class of autoantigens of anti-neutrophil cytoplasmic antibodies in necrotizing and crescentic glomerulonephritis: The lysosomal membrane glycoprotein h-lamp-2 in neutrophil granulocytes and a related membrane protein in glomerular endothelial cells. J Exp Med 1995; 181:585-597.
91. Pierre P, Scheel J, Rickard JE et al. CLIP-170 links endocytic vesicles to microtubules. Cell 1992; 70:887-900.
92. Griffith KJ, Ryan JP, Senecal JL et al. The cytoplasmic linker protein CLIP-170 is a human autoantigen. Clin Exp Immunol 2002; 127:533-538.

CHAPTER 11

Endosomes—Key Components in Viral Entry and Replication

Mark Marsh*

Abstract

Endosomes play key roles in the cellular infection cycles of many viruses. Initially implicated in virus entry, recent research has demonstrated that endosomes can also be required at other stages in viral replication. Endosomes can provide platforms for viral nucleic acid replication and virus assembly, or play roles in modulating anti-viral immune responses. To these ends viruses exploit various attributes of endosomes such as the low luminal pH, unique trafficking properties, cellular location and composition. In turn, viruses have become remarkable tools for analysing endosome function.

Introduction

Endosomes were initially described as intermediates between the plasma membrane and lysosomes in the endocytic transport pathway.[1] Increasingly, these organelles are being recognised as key regulators of endocytic membrane traffic, protein sorting and signalling and they play essential roles in a multitude of cellular functions. Endosomes also have key functions in the activities of many pathogens and toxins. For viruses, endosomes were initially implicated in entry, but recent research has found that endosomes can also function at other stages in viral replication. Endosomes can provide platforms for viral nucleic acid replication and virus assembly, or play roles in modulating anti-viral immune responses. To these ends, viruses exploit the properties of endosomes, including the low luminal pH, trafficking functions, cellular location and composition. In turn, viruses have become remarkable tools for analysing endosome function. I will discuss a number of these activities focussing primarily on the roles of endosomes in virus entry and assembly.

Virus Entry

Prior to the early 1980's, electron microscopic (EM) studies had suggested a role for endocytosis in the cellular entry and infection by a number of viruses.[2] However, knowledge of the pathways involved was at best rudimentary and the molecular mechanisms were not understood. Insights to how endocytosis and endosomes could be used in the entry of cell-free viruses initially came from work on several enveloped viruses, especially Semliki Forest virus (SFV).[3] Indeed one of the original lines of evidence that lead to the description of endosomes as discrete components of the endocytic pathway was from studies of SFV entry.[4] Importantly, this work established that viral endocytosis is not just a cell's attempt to send a potentially

*Corresponding Author: Mark Marsh—Cell Biology Unit, MRC-Laboratory for Molecular Cell Biology, and DeEpartment of Biochemistry and Molecular Biology, University College London, Gower Street, London WC1E 6BT, U.K. Email: m.marsh@ucl.ac.uk

Endosomes, edited by Ivan Dikic. ©2006 Landes Bioscience and Springer Science+Business Media.

harmful agent to lysosomes, but is an essential part of the mechanism used by many viruses to infect cells. These conclusions were reached by applying a range of experimental approaches including biochemistry, morphology, infectivity and pharmacological approaches to attempt to understand virus entry. Much of this work has been reviewed extensively.[5-8]

Cell-free viruses protect their genetic material within a protein shell (nonenveloped viruses), or a protein shell surrounded by a lipid membrane (enveloped viruses). To infect a cell, a virus must dock to receptors on the surface of the target cell, release its DNA or RNA from within the coat (uncoating) and transfer it across a limiting membrane of the cell to the cytoplasm (penetration). Uncoating and penetration are usually tightly coupled to ensure that the viral nucleic acid is only released when there is also the potential for it to reach the host cell cytoplasm.

Three types of mechanism to regulate uncoating and penetration have been identified to date. These can operate on both enveloped and nonenveloped viruses and two require the functional activities of endosomes. In the first, docking of the viruses to specific receptors can trigger uncoating and penetration, i.e., the receptors act as keys to unlock the virus. One of the best-characterised examples of this is the entry of the human immunodeficiency virus (HIV), an enveloped retrovirus. HIV binds to a binary receptor complex of CD4 together with CCR5 or CXCR4 (seven transmembrane domain G protein-coupled receptors for chemokines). Binding of the viral envelope protein (Env) to CD4 initiates conformational changes in Env that generate a second binding site specific for the chemokine receptor. Binding to the chemokine receptor initiates further conformational changes in Env that expose the fusion peptide in the gp41 (transmembrane) subunit of Env, resulting in fusion of the viral membrane with the target cell plasma membrane (see ref. 9). The fusion event simultaneously releases the viral RNA-containing core from within the virion to the cytoplasm. A similar type of receptor-driven uncoating/penetration reaction is believed to occur for poliovirus, a nonenveloped picornavirus. In this case, binding of the poliovirus receptor (PVR/CD155) to 'canyons' that surround protrusions on the surface of the virus particle loosens the interactions between the protein subunits of the virus capsid. This allows the internal capsid protein VP4 and the N-terminus of VP1 to be exposed on the virus surface and insert into the target cell membrane. This is believed to result in the formation of a pore through which the viral RNA is delivered to the cytoplasm.[10,11] Poliovirus and HIV bind their receptors at the cell surface, and penetration probably occurs at the plasma membrane. However for other picornaviruses and retroviruses (and perhaps for poliovirus as well), penetration may also occur in endosomes following endocytosis from the cell surface.[12-14] Indeed the endosomal route may offer some distinct advantages (see below).

The second mechanism for regulating uncoating and penetration is via pH changes. Rather that being receptor-driven, the events leading to the fusion of enveloped viruses or uncoating/penetration of nonenveloped viruses are triggered by exposure of the virions to mildly acidic pH. One of the best-characterised examples of this type of mechanism is the enveloped orthomyxovirus, influenza. The fusogenic envelope protein of influenza virus is the haemmagglutinin (HA). HA binds specifically to sialic acid-containing plasma membrane glycoproteins and glycolipids, which mediate endocytosis of intact virus particles. Following delivery to endosomes, the low pH triggers conformational changes in HA that expose the fusion peptide and initiate membrane fusion (reviewed in ref. 15). Many other enveloped viruses also undergo pH-dependent fusion.[15] In addition, as with the receptor-driven reactions, pH-dependent uncoating and penetration can occur with nonenveloped viruses, including the adenoviruses and some picornaviruses. However, the molecular mechanisms involved in the penetration of these viruses remain to be established in detail. The dependency of these viruses on exposure to low pH requires that they undergo endocytosis to reach acidic early and/or late endosomes.

A third mechanism for regulating entry has emerged from studies of nonenveloped mammalian reoviruses. Following endocytosis of reovirus particles, endosomal cathepsins L and B cleave the σ3 and µ1/µ1C outer-capsid proteins to generate infectious sub-viral particles capable of mediating penetration (reviewed by ref. 16). A similar type of mechanism also appears to apply for the enveloped filovirus, Ebola. Entry is dependent on membrane fusion catalysed by the viral envelope glycoprotein (GP). Earlier work established that Ebola virus is pH-dependent for entry[17,18] but, in contrast to viruses such as influenza and SFV, exposure to low pH does not induce GP membrane fusion activity.[19] It appears that cleavage of GP by a pH-dependent endosomal proteases, probably cathepsin B, or to some extent cathepsin L, is required to trigger fusion.[20] Again, the pH-dependency and proteolytic requirements of this virus require that it must undergo endocytosis for infection.

For acid-dependent viruses, endosomes play an essential role in entry and productive infection. Thus agents that raise the pH of endosomes (weak bases, carboxylic ionophores and specific inhibitors of the vacuolar H^+-ATPase) can inhibit entry and productive infection.[21,22] Depending on the pH required to trigger fusion, different endosome sub-compartments may be used. Hence wt SFV, which undergoes fusion at ~pH 6.2 enters cells from early endosomes,[23,24] whereas strains of influenza, which fuse at ~pH 5.5, penetrate from late endosomes.[25,26] In the absence of endocytosis and delivery to these compartments, infection does not occur. Although essential for the entry of acid-dependent viruses, endosomes may also be used by pH-independent viruses that do not require exposure to acidic pH. These viruses can be internalised by endocytosis and, assuming that the molecular components necessary to trigger the fusion/penetration reactions are present in endosomes, these viruses may also use the endosomal entry route.

Virus Endocytosis

SFV is internalised very efficiently through clathrin-coated vesicles (CCVs) and is delivered from these vesicles to early endosomes.[3,4,27] Experiments using antibodies and dominant negative constructs targeted at proteins involved in clathrin-mediated endocytosis (e.g., dynamin, Eps-15), indicated that clathrin-mediated endocytosis is essential for infection by SFV and related viruses.[28-30] However, these and similar studies also indicated that not all pH-dependent viruses use CCVs. Influenza virus, for example, can undergo endocytosis through CCVs, but inhibition of clathrin-mediated endocytosis does not significantly diminish infectivity, indicating that a clathrin-independent endocytic mechanism can also mediate internalisation of this virus.[30] At least three distinct clathrin-independent pathways have now been identified using different enveloped and nonenveloped viruses, at least two of these pathways appear to involve lipid-raft dependent processes and one of these involves the protein caveolin.[8] This latter pathway is especially important for the entry of simian virus 40 (SV40), a pH-independent, nonenveloped, polyomavirus,[31] though a raft-dependent pathway can also mediate uptake of this virus in caveolin-1 negative cells.[32] These experiments illuminate the multiplicity and versatility of endocytic mechanisms in animal cells and the ability of viruses to capitalise on these activities.

Caveosomes—Parallel Endosome-like Organelles

Following uptake in caveolae, SV40 particles are delivered to endosome-like organelles that have been termed caveosomes. SV40 does not undergo penetration in caveosomes but is sorted into membrane-bound tubules that transfer the virus to the endoplasmic reticulum where, through processes that are not currently understood, it is transferred to the cytoplasm and subsequently to nuclear pores, through which it enters the nucleus.[31] Caveosomes appear not to undergo acidification, to have stable caveolin-coated membrane domains and high cholesterol content. They also appear to be able to undergo a Rab5-dependent interaction with conventional endosomes that may be visualised readily in cells over-expressing dominant active Rab5.[33] In addition to providing a route for virus entry, these organelles have been implicated in the cycling and turnover of lipid raft domains and membrane components enriched in lipid rafts, in particular GPI-linked proteins. However, their precise function remains unclear.[34] For the remainder of this review the term 'endosomes' will be used to describe the acidic, prelysosomal, organelles that receive membrane and content from CCVs.

Endosome Function in Virus Entry

The endosomal route of entry is frequently considered purely a mechanism to provide acid-dependent viruses with a portal into the cell. However, the endocytic pathway leads to the noxious environment of the lysosome, so why have some viruses evolved to use this risky route? Use of endocytic mechanisms ensures that viruses only infect viable cells with a functional endocytic pathway (and not mammalian red blood cells, for example) and allows the virus to control when and where it undergoes penetration. Uptake in endocytic vesicles may allow virus particles to pass through the cortical cytoskeleton that supports the plasma membrane,[35] and transport within endosomes may deliver viruses to specific cellular sites that are crucial for their replication. The unassisted movement of large molecular complexes in the cytoplasm is inefficient. Endosomes can recruit motor proteins such as dynein and move on cytoplasmic microtubules towards the microtubule organising centre (MTOC).[36] Viruses can exploit this activity to hitch a ride to the centre of the cell without the need to carry their own mechanisms for interacting with transport machineries.[37] Thus viruses that replicate in the nucleus, such as influenza and adenovirus, may infect cells more efficiently when carried to the perinuclear region in endosomes. By requiring the more acidic conditions found in late endosomes, viruses may bias their chances of fusing or penetrating from endosomes positioned close to the nucleus.[26] These types of transport events may be especially important for some infection events in vivo, where cell architecture tends to be more developed than in many culture models.

For pH-dependent enveloped viruses, the fusion events that lead to productive infection have usually been considered to occur at the limiting membrane of endosomes. Conceptually, fusion could also occur with the internal membranes of multivesicular endosomes, but there has been little indication that such fusion events could lead to productive infection. Recent studies, primarily with the rhabdovirus vesicular stomatitis virus (VSV), have suggested that such a pathway may be used in some circumstances.[38] This notion stems from experiments in which interference with transport from early to late endosomes appeared to inhibit virus entry, even though virus could be shown to fuse in the early endosome compartment. One interpretation of these studies was that viral fusion occurred with the internal vesicles in multivesicular endosomes (also called multivesicular bodies [MVB]), however, the viral RNA does not enter the cytoplasm until these internal vesicles reach late endosomes where they may fuse with the limiting membrane. Currently, it remains unclear whether this mode of entry is peculiar to the tissue culture model, why even in this model virus should prefer to fuse with the internal membranes and how the fusion of vesicles back to the limiting membrane occurs. However, this pathway resembles a mechanism proposed for the entry of anthrax toxin into cells, where again interaction with the internal membranes of MVBs appears to be crucial.[39] Moreover, influenza virus penetration from late endosomes appears to be blocked when cycling of the ESCRT machinery (see below) is inhibited, suggesting that normal MVB formation and function is also required for penetration by this virus.[40]

For alphaviruses, the limiting membranes of endosomes and lysosomes provide platforms for RNA replication. EM studies of alphavirus-infected cells show characteristic vacuolar structures termed 'cytopathic vacuoles' (CPV) similar in size to endosomes and lysosomes. The prominent feature of these CPV is that their surface is lined with 50 nm diameter vesicular invaginations, or spherules, that project into the lumen of the vacuole. These structures have been shown to be sites of viral RNA synthesis and to contain viral nonstructural proteins (reviewed by ref. 41). The spherules might be formed directly following endosomal fusion of incoming viruses, but similar structures are seen in cells transfected with viral genomic RNA, arguing that there is also some capacity for viral components to target endosome/lysosome membranes following their synthesis. The assembly of replication sites may provide a structural framework to enhance RNA synthesis and/or provide some protection from innate defence mechanisms that target viral RNA.[41]

Endosomes in Virus Assembly

The endocytic pathway provides a mechanism for regulating the cell surface levels of many membrane proteins, in particular signal transducing surface receptors that influence cell division, differentiation and function. In a number of well-studied cases it appears that, following agonist binding and activation, mono-ubiquitination of the cytoplasmic domain of a receptor molecule by a specific ubiquitin (Ub)-E3 ligase leads to internalisation and sorting of the protein to late endosomes and lysosomes. A key step in this pathway occurs in endosomes, where the ubiquitinated proteins are recognised by a set of protein complexes, termed the ESCRTs (endosomal sorting complex required for transport) (reviewed in refs. 42,43), that sort the Ub-tagged proteins into the membrane vesicles that bud into the lumen of endosomes or MVB. A key recent discovery is that this ESCRT machinery is also required for the topologically related process of enveloped virus assembly. Currently, four genera of RNA viruses including retroviruses, rhabdoviruses, filoviruses and arenaviruses (e.g., Lasser fever virus) have been found to require the ESCRT machinery for key steps in their assembly.[42]

The role of the ESCRTs has been most intensively studied for retroviruses. For these viruses, assembly and budding is directed primarily by Gag, the polyprotein precursor of the matrix, capsid and nucleocapsid proteins. Retroviral Gag proteins all contain so-called late domains (L-domains) in which specific amino acid motifs are required for completing the budding reactions.[44-46] To date three distinct types of L-domain motif have been identified, exemplified by (i) the PTAP motif in the HIV Gag L-domain (p6), (ii) the PPxY motif in murine leukaemia virus (MLV) and (iii) the YxxL motif in an equine lentivirus, equine infectious anaemia virus (EIAV) (reviewed by ref. 42). Although quite distinct, these motifs can be moved within the Gag sequence and can be swapped between viruses without compromising assembly, suggesting they have similar functional activities. Significantly, each of the motifs has been found to interact with distinct ESCRT components. The PTAP motif mimics a PSAP sequence in the cellular protein Hrs, which recruits ESCRT-1 to endosomes through interaction with Tsg-101.[47,48] PPxY motifs are consensus binding signals for WW domain containing proteins. Recent studies indicate that MLV Gag binds specifically to WWP1, WWP2 and ITCH, a subgroup of WW and HECT domain-containing, Nedd4-related E3 ligases.[49] These proteins have been implicated in the ubiquitination events described above and may interact with ESCRT complexes.[50,51] Finally, the YxxL motif interacts with AIP1/ALIX,[52] another ESCRT-associated protein that has recently been implicated in the inward invagination of membrane vesicles into endosomes.[53] Thus, L-domain motifs can recruit ESCRT components for interaction with Gag assemblies.

HIV normally buds at the surface of infected T cells. However, the requirement for ESCRT complexes in HIV assembly, together with data indicating that HIV and SIV contain highly conserved endocytosis signals that limit expression of Env at the cell surface and target Env to endocytic organelles, prompted us to investigate whether HIV may in some situation bud into endosomes.[54-56] HIV particles have occasionally been observed within intracellular vesicles particularly in macrophages.[57] However, the nature of these vesicles and the biological significance of the intracellular pool of virus had not been examined. Recent studies have shown that in primary human monocyte-derived macrophages, HIV assembles primarily on endosomal membranes and that virus accumulates within these organelles (Fig. 1).[58,59] Biochemical analyses of the membrane protein composition of HIV particles released into the medium of these cells show that most of the infectious virus derives from endosomes.[58] This intracellular assembly contrasts with HIV budding at the plasma membrane of T cells, and it remains unclear how viral components are targeted differently in different cellular backgrounds. One possibility is that the kinetics of membrane traffic through the endocytic system differs between macrophages and T cells, with the result that the viral components required to form infectious virions accumulate at the plasma membrane in T cells but on endosomes in

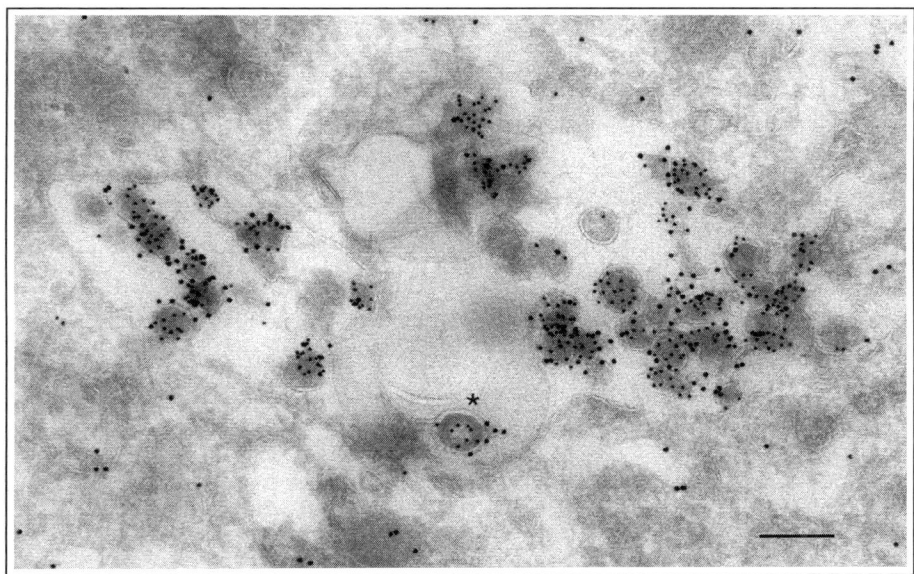

Figure 1. HIV assembly in macrophage endosomes. Human monocyte-derived macrophages were infected with HIV-1$_{Ba-L}$ for 7 days. The cells were then fixed, and ultrathin cryosections prepared for immunolabelling. The figure shows HIV particles within an endosomal vesicle. The virus particles are identified with antibodies against the viral capsid protein (p24) and 10 nm protein A gold. The sections were also labeled for the cellular antigen CD63 (15 nm protein A gold). CD63 can be seen on virus particles, on other membranes in the endosome and on neighboring vesicles. The asterisk identifies an immature virion in which the Gag protein has not yet undergone proteolytic cleavage. Scale bar = 200 nm.

macrophages. Although most (if not all) HIV appears to assemble intracellularly in macrophages, this virus can nevertheless be released from cells apparently through a secretory mechanism in which virus-containing endosomes fuse with the plasma membrane.[58,59] In some respects this discharge of virus has features in common with the release of the late endosomal/lysosomal cytotoxic granules of CD8 T cells. However, it remains to be seen whether the release of HIV-containing endosomes is regulated (see below).

One of the most prominent cellular proteins incorporated into HIV particles in macrophages is the tetraspanin CD63. This antigen can be seen on viruses in endosomes and is present in the membrane of viruses released into the macrophage medium.[58] CD63 traffics over the plasma membrane and is internalised through a C terminal YxxØ type signal but, in many cells, a major fraction of CD63 is seen in late endosomes and is frequently associated with the internal membranes of MVB.[60] The presence of CD63, together with other late endosomal antigens including LAMP-1 and to a lesser extent LAMP-2 has led to the suggestion that HIV might use late endosomes for assembly in macrophages. Significantly, this compartment is also the MHC class II compartment (see below). However, studies with other tetraspanins have identified another endosomal compartment distinct from the CD63-containing late endosomes that may be the site for virus assembly. The properties of this compartment appear to be modified by the virus infection, most notably by the acquisition of CD63. The exact nature of this novel endosome compartment remains to be established. However, initial studies suggest it may be related to an endocytic compartment in dendritic cells (DC) in which HIV can be sequestered during the process of trans infection of T cells (see below). Thus, for HIV in macrophages, budding into endosomes may have multiple advantages, enabling the virus to

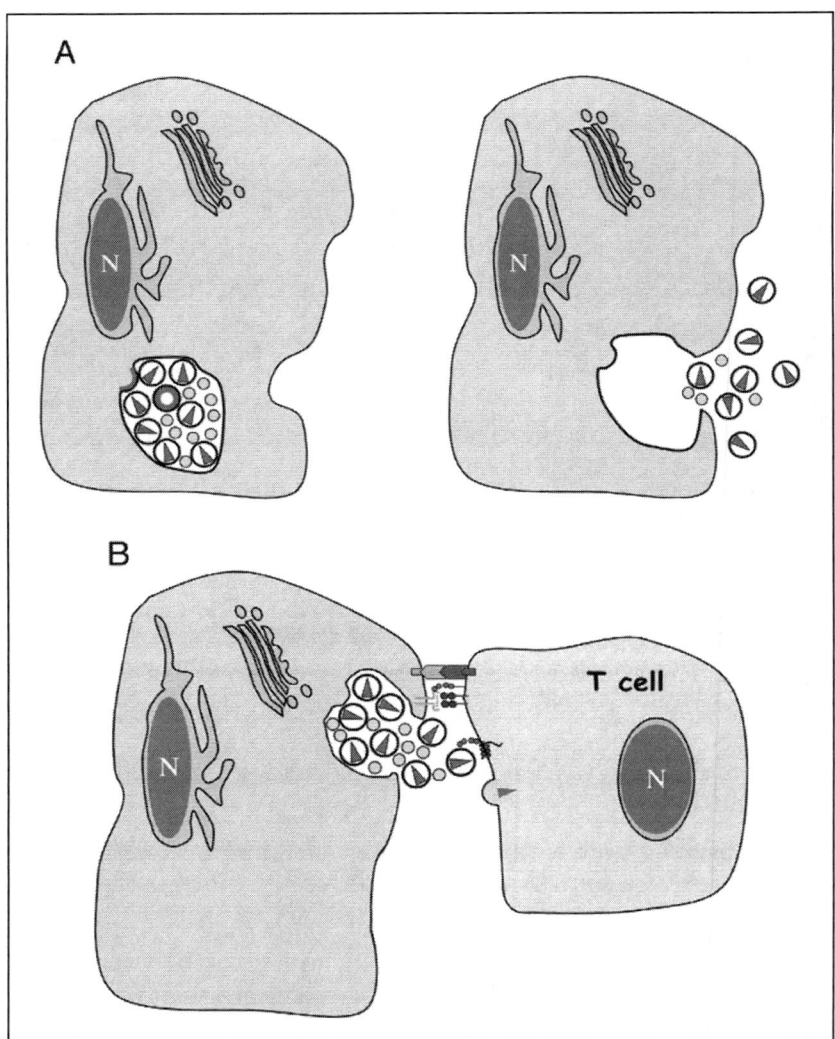

Figure 2. Model for HIV transmission between infected macrophages to T cells. In infected macrophages HIV assembles primarily in an endosomal compartment. This compartment has the capacity to translocate to, and fuse with, the plasma membrane releasing the cargo of viruses (Panel A). Interaction between infected macrophages and T cells may target the virus-containing endosomes to the zones of interaction, which then function as infectious synapses (Panel B). Similar events are proposed to occur when dendritic cells that have acquired virus by endocytosis interact with T cells.

assemble within a sheltered environment hidden from the attentions of the immune system, and allowing virus release to be coupled to the interaction of macrophages with CD4[+ve] T cells—a major host cell for the virus (Fig. 2).[61,62]

Assembly in endosomes appears not to be unique to HIV, as other retroviruses have been seen to bud into endosomes.[63,64] The recycling activities of the endosomal system are required to complete the assembly and release of Mason-Pfizer monkey virus, a B type retrovirus.[65]

Filoviruses have also been reported to require endosomal trafficking to target viral components to the cell surface for assembly,[66] and the β herpesvirus, human cytomegalovirus (HCMV), can bud into MVB and smaller vesicles and membrane cisternae related to late endosomes/MVB.[67] As with HIV, CD63 can be found in the HCMV envelope and, given that this virus can infect macrophages, the possibility exists that it has also evolved to exploit the endocytic system to facilitate its cell-to-cell transfer.

Endosomes in Antigen Presentation

Endosomes play a key role in antigen acquisition, processing and presentation on major histocompatibility type II (MHC-II) antigens. Dendritic cells (DC) patrol surface tissues to monitor for infection. These cells are particularly potent antigen presenting cells (APC) able to capture and present antigens to T cells. In immature DC (iDC), MHC-II antigens are stored within endosomal populations, the so-called MHC II compartment, or MIICs, in a nonpeptide bound form. EM shows this intracellular MHC-II to be located on vesicles within the MIICs.[68] However, when iDC are exposed to antigen and/or activation signals, such as bacterial lipopolysaccharide (LPS), they undergo a maturation programme during which the cells migrate to lymphoid organs and MHC-II molecules are loaded with peptide antigens and presented on the cell surface. During this process the proteins are loaded with antigen that can then be presented to CD4 positive T cells. Viral antigens derived from exogenous viruses taken up by endocytosis can be presented through this mechanism. Such a process may be particularly relevant to HIV where, as discussed above, budding may occur into endosomes. Indeed, in HIV-infected APCs, presentation of HIV antigens to HIV-specific T cells may result in efficient infection of these cells, and may account for the preferential loss of HIV-specific CD4 positive T cells early in the course of infection and abrogation of HIV-directed immune responses.[69]

Endosomal compartments in APCs may also play a role in disseminating viruses. For HIV, DCs are able to transmit virus to CD4^{+ve} T cells without themselves becoming infected. This process of trans infection was first reported by Steinman et al who found that the most effective way of infecting T cells with HIV in culture was to present this virus in the context of a DC.[70] Recent experiments have begun to shed some light on the mechanisms underlying these events. APCs express a range of receptors able to bind and internalise biochemically distinct antigens. One group of these receptors is the C type lectins with specificity for high mannose oligosaccharides. Several of these lectins, in particular DC-SIGN (dendritic cell-specific ICAM-3 grabbing nonintegrin), can bind HIV Env without triggering fusion or infection.[71] The bound viruses are internalised, in part by clathrin-mediated endocytosis, and delivered to an endosomal compartment with properties distinct from conventional early and late endosomes.[72,73] Morphological observations indicate these organelles have internal membranes, i.e., they are multivesicular, and they contain a number of tetraspannins including CD9 and CD81, similar to the macrophage compartment where HIV assembles. When DC are then incubated with T cells, the virus-containing vesicles migrate to the zone of interaction between the DC and T cell and virus is transferred to the T cell, which becomes efficiently infected.[73,74] The zone of DC-T cell interaction through which virus transmission occurs has properties in common with immunological synapses and has been termed the 'infectious' or 'virological' synapse. These synapses may represent a mechanism that viruses have developed to exploit the special signalling activities of immunological synapses to facilitate their transfer from infected to uninfected cells. It is possible that HIV targets a special endosome compartment in APC (in particular macrophages and DC) either by endocytic uptake, or by assembly if the cell is infected, and that this compartment is targeted to the APC-T cell synapses.

Infectious synapses have been seen between infected and uninfected T cells for HIV, and for HTLV-1, where roles for endosomes are unclear. But the fact that a number of viral pathogens, including SIV,[75] HCMV,[76] hepatitis C virus[77] and Ebola virus,[78] interact with DC through DC-SIGN and/or other C-type lectins, suggest that a number of viruses may traffic through DC endosomes to enhance infection.

Viral Modulation of Endosome Trafficking

Viruses also exploit endocytic membrane trafficking pathways to modulate the distribution of cellular proteins important for pathogenesis. For example, in HIV infected cells plasma membrane CD4 is down-modulated by the viral Nef protein. This small myristoylated peripheral membrane protein can act as a linker to couple proteins that may not normally interact. Nef can induce the association of CD4 with the clathrin AP2 adaptor complex to bring about its uptake from the cell surface. In endosomes, Nef also appears to link CD4 to an endosomal complement of the COP-1 coat proteins thereby directing internalised CD4 molecules to lysosomes.[79,80] Similar mechanisms may regulate the levels of other cell surface proteins in HIV infected cells, though the modulation of major histocompatibility type I (MHC-I) antigens in HIV infected T cells appears to involve retargeting of newly synthesised MHC-I through Nef interaction with AP1 adaptors during transport through the exocytic pathway.[81]

Many viruses are able to down-modulate MHC-I expression to abrogate detection of infected cells by cytotoxic T lymphocytes. In a number of cases this involves perturbation of MHC-I synthesis but in others endocytic trafficking is required. KSHV, a γ herpesvirus linked to Karposi's sarcoma, appears to use mono-ubiquitin to target MHC-I antigens to lysosomes. This virus encodes two E3 ligases (modulator of immune recognition [MIR]1 and MIR2) that mono-ubiquitinate MHC-I heavy chains to induce their internalisation and sorting through endosomes to lysosomes (see above).[82] Other antigens, including ICAM-1, may be modulated by a similar mechanism. By contrast, the murine cytomegalovirus uses the gp48 protein encoded by the early gene m06 to bind newly synthesised murine MHC-I and, through an endosomal di-leucine trafficking motif, target the MHC-I antigens to endosomes and lysosomes for degradation.[83] HCMV employs ER retention mechanisms to reduce MHC-I cell surface expression, but this virus may use additional mechanisms to hide infected cells. As with many other herpesviruses, HCMV encodes seven transmembrane domain proteins that share features with chemokine receptors. One such protein is US28, which binds a range of CC and CXC chemokines, as well as the unusually membrane-bound CX_3C chemokine—fractalkine. By contrast to many cellular chemokine receptors, US28 is constitutively active for endocytosis and recycling[84] and may function as a chemokine sink, removing these chemoattractants from the environment around infected cells by endocytosis.[85]

As detailed studies of different viruses continue, it is likely that other mechanisms will emerge through which alteration in protein trafficking through endosomes play a role in modulating the expression of MHC and other cellular antigens.

Conclusions

In one way and another, endosomes have been exploited by a large number of viruses to facilitate their replication. The role of endosomes in the entry of many enveloped and nonenveloped viruses is now well established, although in many cases the details of these events still have to be worked out. For many RNA viruses, endosomes can provide platforms for the replication of viral nucleic acid and endosomes may be manipulated by the virus to provide a sanctuary in which replication can occur in isolation from cellular defence mechanisms. The nature of the immune response to viruses can also be manipulated through exploiting the properties of endosomes. Finally, and perhaps most surprisingly, some viruses appear to use the endocytic system for assembly. Given the close links between endosomes and lysosomes, this might seem to be a risky strategy but also suggests it has significant benefits for the virus. It may be that endosomal assembly is a strategy used most effectively in APC, where the endocytic pathway is modified to facilitate antigen presentation, and the viruses can exploit these properties to enhance their cell-to-cell transmission.

Over the last 20 years, work with viral systems has provided excellent tools for analysing the properties of endosomes. There is still much to learn and it is likely that viruses will continue to provide insights into the functions of these organelles. Moreover, it is likely that further novel mechanisms through which viruses exploit the properties of endosomes to facilitate their replication and transmission will be discovered.

Acknowledgements

Work in the author's laboratory is supported by the UK Medical Research Council and grant number AI-49784 from the National Institutes of Health. I thank Annegret Pelchen-Matthews for critical comments on the manuscript.

References

1. Helenius A, Mellman I, Wall D et al. Endosomes. Trends Biochem Sci 1983; 8(7):245.
2. Dales S. Early events in cell-animal virus interactions. Bacteriol Rev 1973; 37:103-135.
3. Helenius A, Kartenbeck J, Simons K et al. On the entry of Semliki Forest virus into BHK-21 cells. J Cell Biol 1980; 84:404-420.
4. Marsh M, Bolzau E, Helenius A. Semliki Forest virus penetration occurs from acid prelysosomal vacuoles. Cell 1983; 32:931-940.
5. Marsh M, Helenius A. Virus entry into animal cells. Adv Virus Res 1989; 36:107-151.
6. Smith AE, Helenius A. How viruses enter animal cells. Science 2004; 304(5668):237-42.
7. Sieczkarski SB, Whittaker GR. Viral entry. Curr Top Microbiol Immunol 2005; 285:1-23.
8. Pelkmans L, Helenius A. Insider information: What viruses tell us about endocytosis. Curr Opin Cell Biol 2003; 15(4):414-22.
9. Chen B, Vogan EM, Gong H et al. Structure of an unliganded simian immunodeficiency virus gp120 core. Nature 2005; 433(7028):834.
10. Belnap DM, Filman DJ, Trus BL et al. Molecular tectonic model of virus structural transitions: The putative cell entry states of poliovirus. J Virol 2000; 74(3):1342-54.
11. Belnap DM, McDermott Jr BM, Filman DJ et al. Three-dimensional structure of poliovirus receptor bound to poliovirus. Proc Natl Acad Sci USA 2000; 97(1):73-8.
12. Bayer N, Prchla E, Schwab M et al. Human rhinovirus HRV14 uncoats from early endosomes in the presence of bafilomycin. FEBS Lett 1999; 463(1-2):175-8.
13. Schober D, Kronenberger P, Prchla E et al. Major and minor receptor group human rhinoviruses penetrate from endosomes by different mechanisms. J Virol 1998; 72(2):1354-64.
14. Bayer N, Schober D, Prchla E et al. Effect of bafilomycin A1 and nocodazole on endocytic transport in HeLa cells: Implications for viral uncoating and infection. J Virol 1998; 72(12):9645-55.
15. Earp LJ, Delos SE, Park HE et al. The many mechanisms of viral membrane fusion proteins. Curr Top Microbiol Immunol 2005; 285:25-66.
16. Ebert DH, Deussing J, Peters C et al. Cathepsin L and cathepsin B mediate reovirus disassembly in murine fibroblast cells. J Biol Chem 2002; 277(27):24609-17.
17. Wool-Lewis RJ, Bates P. Characterization of Ebola virus entry by using pseudotyped viruses: Identification of receptor-deficient cell lines. J Virol 1998; 72(4):3155-60.
18. Takada A, Robison C, Goto H et al. A system for functional analysis of Ebola virus glycoprotein. Proc Natl Acad Sci USA 1997; 94(26):14764-9.
19. Ito H, Watanabe S, Sanchez A et al. Mutational analysis of the putative fusion domain of Ebola virus glycoprotein. J Virol 1999; 73(10):8907-12.
20. Chandran K, Sullivan NJ, Felbor U et al. Endosomal proteolysis of the Ebola virus glycoprotein is necessary for infection. Science 2005; 308(5728):1643-5.
21. Helenius A, Marsh M, White J. Effect of lysosomotropic weak bases on Semliki Forest virus penetration into host cells. J Gen Virol 1982; 58:47-61.
22. Marsh M, Wellsteed J, Kern H et al. Monensin inhibits Semliki Forest virus penetration into culture cells. Proc Natl Acad Sci USA 1982; 79:5297-5301.
23. Kielian MC, Marsh M, Helenius A. Kinetics of endosome acidification detected by mutant and wild-type Semliki Forest virus. EMBO J 1986; 5:3103-3109.
24. Schmid S, Fuchs R, Kielian M et al. Acidification of endosome subpopulations in wild-type Chinese hamster ovary cells and temperaturesensitive acidification-defective mutants. J Cell Biol 1989; 108(4):1291-1300.
25. Sieczkarski SB, Whittaker GR. Differential requirements of Rab5 and Rab7 for endocytosis of influenza and other enveloped viruses. Traffic 2003; 4(5):333-343.
26. Lakadamyali M, Rust MJ, Babcock HP et al. Visualizing infection of individual influenza viruses. Proc Natl Acad Sci USA 2003; 100(16):9280-5.
27. Marsh M, Helenius A. Adsorptive endocytosis of Semliki Forest virus. J Mol Biol 1980; 142:439-454.

28. DeTulleo L, Kirchhausen T. The clathrin endocytic pathway in viral infection. EMBO J 1998; 17(16):4585-93.
29. Doxsey SJ, Brodsky FM, Blank GS et al. Inhibition of endocytosis by anti-clathrin antibodies. Cell 1987; 50:453-463.
30. Sieczkarski SB, Whittaker GR. Influenza virus can enter and infect cells in the absence of clathrin-mediated endocytosis. J Virol 2002; 76(20):10455-64.
31. Pelkmans L, Kartenbeck J, Helenius A. Caveolar endocytosis of simian virus 40 reveals a new two-step vesicular-transport pathway to the ER. Nat Cell Biol 2001; 3(5):473-83.
32. Damm EM, Pelkmans L, Kartenbeck J et al. Clathrin- and caveolin-1-independent endocytosis: Entry of simian virus 40 into cells devoid of caveolae. J Cell Biol 2005; 168(3):477-88.
33. Pelkmans L, Burli T, Zerial M et al. Caveolin-stabilized membrane domains as multifunctional transport and sorting devices in endocytic membrane traffic. Cell 2004; 118(6):767-80.
34. Pelkmans L, Helenius A. Endocytosis via caveolae. Traffic 2002; 3(5):311-20.
35. Marsh M, Bron R. SFV infection in CHO cells: Cell-type specific restrictions to productive virus entry at the cell surface. J Cell Science 1997; 110:95-103.
36. Valetti C, Wetzel DM, Schrader M et al. Role of dynactin in endocytic traffic: Effects of dynamitin overexpression and colocalization with CLIP-170. Mol Biol Cell 1999; 10(12):4107-20.
37. Georgi A, Mottola-Hartshorn C, Warner A et al. Detection of individual fluorescently labeled reovirions in living cells. Proc Natl Acad Sci USA 1990; 87(17):6579-83.
38. Le Blanc I, Luyet PP, Pons V et al. Endosome-to-cytosol transport of viral nucleocapsids. Nat Cell Biol 2005.
39. Abrami L, Lindsay M, Parton RG et al. Membrane insertion of anthrax protective antigen and cytoplasmic delivery of lethal factor occur at different stages of the endocytic pathway. J Cell Biol 2004; 166(5):645-51.
40. Khor R, McElroy LJ, Whittaker GR. The ubiquitin-vacuolar protein sorting system is selectively required during entry of influenza virus into host cells. Traffic 2003; 4(12):857-68.
41. Salonen A, Ahola T, Kaariainen L. Viral RNA replication in association with cellular membranes. Curr Top Microbiol Immunol 2005; 285:139-73.
42. Morita E, Sundquist WI. Retrovirus budding. Annu Rev Cell Dev Biol 2004; 20:395-425.
43. Pelchen-Matthews A, Raposo G, Marsh M. Endosomes, exosomes and trojan viruses. Trends Microbiol 2004; 12(7):310-316.
44. Gottlinger HG, Dorfman T, Sodroski JG et al. Effect of mutations affecting the p6 gag protein on human immunodeficiency virus particle release. Proc Natl Acad Sci USA 1991; 88(8):3195-9.
45. Wills JW, Cameron CE, Wilson CB et al. An assembly domain of the Rous sarcoma virus Gag protein required late in budding. J Virol 1994; 68(10):6605-18.
46. Parent LJ, Bennett RP, Craven RC et al. Positionally independent and exchangeable late budding functions of the Rous sarcoma virus and human immunodeficiency virus Gag proteins. J Virol 1995; 69(9):5455-60.
47. Bache KG, Brech A, Mehlum A et al. Hrs regulates multivesicular body formation via ESCRT recruitment to endosomes. J Cell Biol 2003; 162(3):435-42.
48. Pornillos O, Higginson DS, Stray KM et al. HIV Gag mimics the Tsg101-recruiting activity of the human Hrs protein. J Cell Biol 2003; 162(3):425-34.
49. Martin-Serrano J, Eastman SW, Chung W et al. HECT ubiquitin ligases link viral and cellular PPXY motifs to the vacuolar protein-sorting pathway. J Cell Biol 2005; 168(1):89-101.
50. Segura-Morales C, Pescia C, Chatellard-Causse C et al. Tsg101 and Alix interact with MLV Gag and cooperate with Nedd4 ubiquitin-ligases during budding. J Biol Chem 2005; 280(29):27004-12.
51. Medina G, Zhang Y, Tang Y et al. The functionally exchangeable L domains in RSV and HIV-1 Gag direct particle release through pathways linked by Tsg101. Traffic 2005; 6(10):880-94.
52. Strack B, Calistri A, Craig S et al. AIP1/ALIX is a binding partner for HIV-1 p6 and EIAV p9 functioning in virus budding. Cell 2003; 114(6):689-99.
53. Matsuo H, Chevallier J, Mayran N et al. Role of LBPA and Alix in multivesicular liposome formation and endosome organization. Science 2004; 303(5657):531-4.
54. Rowell JF, Stanhope PE, Siliciano RF. Endocytosis of endogenously synthesized HIV-1 envelope protein. Mechanism and role in processing for association with class II MHC. J Immunol 1995; 155(1):473-88.
55. Sauter MM, Pelchen-Matthews A, Bron R et al. An internalization signal in the simian immunodeficiency virus transmembrane protein cytoplasmic domain modulates expression of envelope glycoproteins on the cell surface. J Cell Biol 1996; 132:795-812.

56. Bowers K, Pelchen-Matthews A, Höning S et al. The simian immunodeficiency virus envelope glycoprotein contains multiple signals that regulate its cell surface expression and endocytosis. Traffic 2000; 1:661-674.

57. Orenstein JM, Meltzer MS, Phipps T et al. Cytoplasmic assembly and accumulation of human immunodeficiency virus types 1 and 2 in recombinant human colony-stimulating factor-1-treated human monocytes: An ultrastructural study. J Virol 1988; 62(8):2578-86.

58. Pelchen-Matthews A, Kramer B, Marsh M. Infectious HIV-1 assembles in late endosomes in primary macrophages. J Cell Biol 2003; 162:443-455.

59. Raposo G, Moore M, Innes D et al. Human macrophages accumulate HIV-1 particles in MHC II compartments. Traffic 2002; 3(10):718-29.

60. Escola JM, Kleijmeer MJ, Stoorvogel W et al. Selective enrichment of tetraspan proteins on the internal vesicles of multivesicular endosomes and on exosomes secreted by human B-lymphocytes. J Biol Chem 1998; 273(32):20121-7.

61. Carr JM, Hocking H, Li P et al. Rapid and efficient cell-to-cell transmission of human immunodeficiency virus infection from monocyte-derived macrophages to peripheral blood lymphocytes. Virology 1999; 265(2):319-29.

62. Sharova N, Swingler C, Sharkey M et al. Macrophages archive HIV-1 virions for dissemination in trans. EMBO J 2005; 24(13):2481-9.

63. Nydegger S, Foti M, Derdowski A et al. HIV-1 egress is gated through late endosomal membranes. Traffic 2003; 4(12):902-10.

64. Sherer NM, Lehmann MJ, Jimenez-Soto LF et al. Visualization of retroviral replication in living cells reveals budding into multivesicular bodies. Traffic 2003; 4(11):785-801.

65. Sfakianos JN, Hunter E. M-PMV capsid transport is mediated by Env/Gag interactions at the pericentriolar recycling endosome. Traffic 2003; 4(10):671-80.

66. Kolesnikova L, Berghofer B, Bamberg S et al. Multivesicular bodies as a platform for formation of the Marburg virus envelope. J Virol 2004; 78(22):12277-87.

67. Fraile-Ramos A, Pelchen-Matthews A, Kledal TN et al. Localization of human cytomegalovirus UL33 and US27 proteins in endocytic compartments and viral membranes. Traffic 2002; 3:218-232.

68. Kleijmeer M, Ramm G, Schuurhuis D et al. Reorganization of multivesicular bodies regulates MHC class II antigen presentation by dendritic cells. J Cell Biol 2001; 155(1):53-63.

69. Janeway Jr CA. The case of the missing CD4s. Curr Biol 1992; 2(7):359-61.

70. Cameron PU, Freudenthal PS, Barker JM et al. Dendritic cells exposed to human immunodeficiency virus type-1 transmit a vigorous cytopathic infection to CD4+ T cells. Science 1992; 257(5068):383-7.

71. Geijtenbeek TB, van Kooyk Y. DC-SIGN: A novel HIV receptor on DCs that mediates HIV-1 transmission. Curr Top Microbiol Immunol 2003; 276:31-54.

72. Kwon DS, Gregorio G, Bitton N et al. DC-SIGN-mediated internalization of HIV is required for trans-enhancement of T cell infection. Immunity 2002; 16(1):135-44.

73. Garcia E, Pion M, Pelchen-Matthews A et al. Trafficking of HIV to the dendritic cell-T cell infectious synapse uses the pathway of tetraspanin sorting to the immunological synapse. Traffic 2005; 6:488-501.

74. McDonald D, Wu L, Bohks SM et al. Recruitment of HIV and its receptors to dendritic cell-T cell junctions. Science 2003; 300(5623):1295-7.

75. Ploquin MJ, Diop OM, Sol-Foulon N et al. DC-SIGN from African green monkeys is expressed in lymph nodes and mediates infection in trans of simian immunodeficiency virus SIVagm. J Virol 2004; 78(2):798-810.

76. Halary F, Amara A, Lortat-Jacob H et al. Human cytomegalovirus binding to DC-SIGN is required for dendritic cell infection and target cell trans-infection. Immunity 2002; 17(5):653-64.

77. Ludwig IS, Lekkerkerker AN, Depla E et al. Hepatitis C virus targets DC-SIGN and L-SIGN to escape lysosomal degradation. J Virol 2004; 78(15):8322-32.

78. Alvarez CP, Lasala F, Carrillo J et al. C-type lectins DC-SIGN and L-SIGN mediate cellular entry by Ebola virus in cis and in trans. J Virol 2002; 76(13):6841-4.

79. Whitney JA, Gomez M, Sheff D et al. Cytoplasmic coat proteins involved in endosome function. Cell 1995; 83(5):703-13.

80. Aniento F, Gu F, Parton RG et al. An endosomal beta COP is involved in the pH-dependent formation of transport vesicles destined for late endosomes. J Cell Biol 1996; 133(1):29-41.

81. Roeth JF, Williams M, Kasper MR et al. HIV-1 Nef disrupts MHC-I trafficking by recruiting AP-1 to the MHC-I cytoplasmic tail. J Cell Biol 2004; 167(5):903-13.

82. Coscoy L, Sanchez DJ, Ganem D. A novel class of herpesvirus-encoded membrane-bound E3 ubiquitin ligases regulates endocytosis of proteins involved in immune recognition. J Cell Biol 2001; 155(7):1265-73.

83. Reusch U, Muranyi W, Lucin P et al. A cytomegalovirus glycoprotein reroutes MHC class I complexes to lysosomes for degradation. EMBO J 1999; 18(4):1081-91.

84. Fraile-Ramos A, Kledal T, Pelchen-Matthews A et al. The human cytomegalovirus US28 protein is located in endocytic vesicles and undergoes constitutive endocytosis and recycling. Mol Biol Cell 2001; 12:1737-1749.

85. Beisser PS, Goh CS, Cohen FE et al. Viral chemokine receptors and chemokines in human cytomegalovirus trafficking and interaction with the immune system. CMV chemokine receptors. Curr Top Microbiol Immunol 2002; 269:203-34.

CHAPTER 12

Toxins in the Endosomes

Núria Reig and F. Gisou van der Goot*

Abstract

Many bacteria owe their virulence to the production of protein toxins. These proteins usually play an important role in permitting the bacteria to successfully spread in the host and cause infection. With the exception of poreforming toxins and lipases, all toxins need to be endocytosed by the host cell to perform their toxin action. In the recent years, the increased knowledge of how toxins enter the cells and reach the cytoplasm have highlighted their ability to exploit, in its finest details, the membrane-trafficking systems of their hosts. In this chapter, based on selected examples we will review how toxins use the host endosomal system to reach their targets.

Introduction

In order to successfully colonize their host, many bacterial pathogens produce protein toxins. These are secreted proteins that modify the behavior of target mammalian cells. Whereas some act at the plasma membrane, such as pore-forming toxins or lipases, most are enzymes with cytoplasmic targets. This implies that they must cross a biological membrane to reach this intracellular milieu. To avoid deleterious membrane permeabilization, toxins never cross the plasma membrane. Instead, they are taken up by cells, transported to intracellular organelles, where membrane translocation occurs. We will here focus on the entry routes of toxins and their trafficking through the endocytic pathway. Five examples have been chosen to illustrate the different ways in which the endocytic pathway can be utilized or altered by toxins: cholera toxin (CT), produced by *Vibrio cholerae* and responsible for the secretory diarrhea associated with cholera disease,[1] Shiga toxin (ST), produced by *Shigella dysenteriae* and responsible for the vascular damage observed in shigellosis,[2] diphtheria toxin (DT), produced by *Corynebacterium diphteriae* and able to kill the intoxicated cells by inhibiting protein synthesis,[3] anthrax toxin, produced by *Bacillus anthracis*, the causative agent of anthrax[9] and Helicobacter vacuolating toxin (vacA), produced by *Helicobacter pylori*, a common colonizer of the human stomach and a risk factor for the development of peptic ulcer disease and gastric adenocarcinoma.[4] All five toxins, with the possible exception of VacA, are formed by 2 subunits, an A subunit that bares the enzymatic activity and a B subunit that has the ability to interact with the host cell and escorts the A subunit to its final destination (Table 1). We will only focus on the interaction of these AB toxins with the endocytic pathways, leaving out important interactions with the biosynthetic pathway for CT and ST and with mitochondria for vacA.

*Corresponding Author: F. Gisou van der Goot—Department of Microbiology and Molecular Medicine, University of Geneva, 1 rue Michel Servet, 1211 Geneva 4, Switzerland, Email: gisou.vandergoot@medecine.unige.ch

Endosomes, edited by Ivan Dikic. ©2006 Landes Bioscience and Springer Science+Business Media.

Table 1.

Toxin	Receptor	Structure	Activity	Target
Anthrax toxin	TEM8, CMG2	3 independent polypeptide chains: EF, LF, PA PA becomes heptameric after contact with the host cell	EF: calmodulin dependent adenylate cyclase, LF: metalloprotease	LF: MAPKKs
Cholera toxin	GM1	2 independant polypeptide chains: A and B, B being pentameric: A-B5	ADP-ribosyltransferase	G-proteins
Diphtheria toxin	HB-EGF	1 single polypeptide chain, with 2 disulfide linked subunits	ADP-ribosyltransferase	EF-2
Shiga toxin	Gb3	2 independant polypeptide chains: A and B, B being pentameric: A-B5	N-glycosidase	28S rRNA
VacA	Unknown	1 single polypeptide chain, with 2 subunits, becomes hexa or heptameric after contact with the host cell	channel forming	plasma membare, late endosomes, mitochondria

Different Portals of Entry into the Cell

Binding of bacterial toxins to targets cells is mediated by the specific interaction of the toxin B subunit, with a cell surface molecule that can be either a sugar, a lipid or a protein (Table 1). The receptor for CT is the ganglioside GM1,[1] for ST the ganglioside Gb3,[2] for DT the heparin-binding epidermal growth factor precursor (HB-EGF)[5] and for anthrax toxin two homologous receptors have been identified, Tumor endothelial marker 8 (TEM8)[6] and Capillary Morphogenesis gene 2 (CMG2).[7] VacA, in contrast to most toxins, appears to bind to a variety of different receptors including Protein-tyrosine phosphatase alpha, various lipids, the epidermal growth factor receptor and heparan sulphate[8] (for review see ref. 4).

Toxin B subunits can be either part of the same polypeptide chain as the A subunit, such as for DT and VacA, or synthesized separately by the bacterium, as for anthrax toxin, CT and ST. In addition, B subunits can be monomeric (DT) or multimeric (pentamers for CT and ST, heptamers for anthrax and VacA). These multimers are formed either during initial folding of the toxin (CT and ST) or at the surface of the target cell (anthrax toxin and VacA). This is of importance for understanding the endocytic routing of these toxins. Preformed multimers are indeed multivalent ligands, thus preferentially targeting regions of the plasma membrane where receptors are clustered. For example CT preferentially binds to caveolae, which contain clusters of GM1. In contrast, B subunits that are produced as monomers by the bacterium are monovalent ligands that, once bound to the cell surface, trigger clustering of the receptors upon oligomerization. This can in turn lead to receptor relocalization as observed for the B subunit of anthrax toxin, which is called the protective antigen.[9,10] It was indeed observed that heptamerization of the protective antigen leads to the redistribution of the receptor from the glycerophospholipidic region of the plasma membrane to specialized lipid domains rich in cholesterol and glycosphingolipids called lipid rafts.[11] Interestingly, four out of the five toxins used as examples in this review associate with lipid rafts during the processes of cell binding and endocytosis: CT,[12] ST,[2] anthrax toxin[11] and VacA.[13] Our own unpublished observations show that DT does not associate with lipid rafts, using detergent resistance as a read out. It was however observed that cells lacking sphingolipids, an important component of lipid rafts as well as other membrane domains, are more sensitive to DT.[14]

Once bound to their respective receptors, toxins are internalized. As described in the previous chapters and recent reviews,[15-17] different pathways, which exist in parallel, allow entry into mammalian cells. These include the well-characterized clathrin-dependent pathway, the more recently characterized caveolar pathway as well as nonclathrin and noncaveolar pathways. As one might expect from opportunistic ligands such as toxins, each of these pathways has been hijacked, and usually one toxin can enter the cell by more than one pathway.

Toxin Entry by Clathrin-Mediated Endocytosis

Of the five example toxins discussed in this review, only vacA appears to be excluded from this pathway.[18] In contrast, for both diphtheria toxin and anthrax toxin, entry via clathrin-coated pits is the preferential route.[11,19,20] Interestingly however, whereas DT enters through constitutive endocytosis of its transmembrane receptor, the anthrax toxin *triggers* receptor uptake, via poorly characterized mechanisms that involve raft association[11] and most probably signaling events. For both toxins, targeting to the clathrin dependent entry route is most likely dependent on motifs in the cytoplasmic tails of their respective receptors, but this has not been analyzed. Surprisingly, ST and CT also can enter cells via clathrin-coated pits despite the fact that they bind to gangliosides, which being in the outer leaflet of the membrane can not interact with cytosolic sorting machineries.[21-26] As for the anthrax toxin, association to lipid rafts might trigger signaling to the endocytic machinery. That Shiga toxin does induce signaling events is illustrated by its activation of the *src*-like kinases *yes* and *lyn*.[27,28] How subsequently adaptors and clathrin triskelions would be recruited to the membrane, or whether the toxin containing rafts would enter preformed pits, is unclear.

Entry via Non-Clathrin Dependent Routes

Of the 5 toxins presented here, most of them can enter the cell by clathrin dependent as well as clathrin independent mechanisms, but only VacA was found to enter exclusively via a clathrin-independent pathway,[18] but the exact pathway has not been further documented other than that it is lipid raft dependent.[13,18,29,30] In contrast, the alternative entry routes were extensively studied for CT.[1,22,23,31,32-34] The common denominator of the clathrin-independent CT entry pathways is their dependence on cellular cholesterol.[12,22,34] Uptake through caveolae has been frequently proposed[1,31,33] however several recent studies have concluded that it is a relatively minor pathway.[22,23,32,34] Using mouse embryonic fibroblasts form caveolin-1 knock out mice, Parton and colleagues[34] have recently shown that CT preferentially enter via two pathways besides the caveolar pathway : the clathrin-coated pit pathway, as mentioned above, and a nonclathrin noncaveolar cholesterol dependent pathway, that is also used by glycosylphosphatidyl inositol (GPI) anchored proteins.[35] Massol et al recently found that although endocytosis was not longer detected, cytotoxicity of CT was retained in cells cotransfected with dominant negative mutants of dynamin (involved in both the clathrin and caveolar pathways) and Arf6 suggesting the existence of a pathway independent of clathrin, caveolin and Arf6.[25] Depending on the cell types, ST was also found to enter cells via a clathrin and dynamin independent pathway,[36,37] but this route was not studied in detail.

Early Endosomes: Toxin Translocation Site or Transit Area

Despite the diversity of the entry requirements and routes, all five toxins reach the early endosomes. The route they have followed might however determine their precise localization within the early endosomes, since this compartment is composed of a mosaic of membrane domains.[38,39] In addition the compartment is polymorphic with tubular and cisternal regions, the later being multivesicular.

For DT, the early endosome is the final destination.[40,41] The acidic milieu of the endosome triggers a conformational change in the B subunit of the toxin that inserts into the endosomal membrane and allows the translocation of the A subunit into the cytoplasm. At that stage, a proteolytic cleavage has occurred between the A and B subunits, which however remain attached via a disulfide bound that becomes reduced by cytosolic factors, possibly thioredoxin reductase.[42] Translocation occurs in an at least partially unfolded state and refolding on the cytoplasmic side is promoted by the cytosolic chaperone HSP90, together with other yet to be identified cytosolic factors.[42] Once in the cytoplasm, the A subunit ADP-ribosylates Elongation Factor 2 (EF2), leading to the inhibition of protein synthesis followed by cell death.[3]

CT and ST are only in transit in early endosomes from where they are directly routed, i.e., without going through later endocytic compartments, to the Trans-Golgi Network (TGN). The association of the gangliosides, GM1 for CT[43] and Gb3 for ST,[44] with detergent resistant membranes, and presumably with lipid rafts, was found to be essential for transport to the TGN. More over transport of ST to the TGN requires clathrin, dynamin, epsinR, and possibly Rab11.[36,37,45] Also two different SNARE complexes, involving syntaxins 5 and 16, were found to be important.[46] This direct retrograde pathway, described for first time for ST, appears to be used by a growing number of Golgi proteins that cycle between the Golgi, the plasma membrane and early endosomes.[47] CT and ST are subsequently transported to the Golgi (Fig. 1) and the endoplasmic reticulum from where they undergo retrograde translocation into the cytoplasm, hijacking the cellular machinery that enables misfolded proteins to cross the ER membrane to reach the cytosol where proteasomal degradation can take place (for review see ref. 48). Once in the cytoplasm, the CT A subunit ADP-ribosylates G proteins leading to the increase of cAMP and subsequent chloride secretion, where as the ST A subunit cleaves a single Adenine base from the 28S ribosomal RNA thus inhibiting protein synthesis.

VacA and anthrax toxin are also in transit in early endosomes on their way to late endosomes. Anthrax toxin was found to require sorting into the cisternal regions of the early endosomes as opposed to the tubular regions involved in recycling.[41] More specifically the toxin is sorted, by yet unknown mechanisms, into nascent intraluminal vesicles of the early endosomes. The biogenesis of intraluminal vesicles is described in Chapter 8 and involves the ubiquination of cargo molecules and the interaction with the three ESCRT (Endosomal sorting complex required for transport) complexes. Once sorted into these vesicles, the heptameric anthrax B subunit inserts into the membrane,[41] in a manner that is dependent on the acidic endosomal pH, and mediates the translocation of the A subunits, of which anthrax toxin has two: lethal factor LF, a metalloprotease that cleaves MAP kinase kinases, and edema factor EF, a calmodulin dependent adenylate cyclase. Since B subunit channel formation occurs in the membrane of intraluminal vesicles, the enzymatic subunits end up in the lumen of these vesicles upon membrane translocation[41] (Fig. 1). Upon budding of multivesicular bodies, or endosomal carrier vesicles (see Chapters 1, 2 and 8), the encapsulated enzymatic toxin subunits are thus withdrawn for the early endosome and transported to late endosomes. As for early to late endosomal transport in general, transport of anthrax toxin A subunits was dependent on the integrity of microtubules.

Late Endosomes: Who Can Make It That Far and What For?

Whereas most, if not all, AB type toxins interact at some point with early endosomes, very few reach later stages of the endocytic pathway. In fact, only two toxins have been described to do so: vacA and anthrax toxin.

The most striking feature of vacA, which led to its name, is that it triggers vacuolation of cells. The compartment undergoing vacuolation was subsequently identified as the late endosome based on its acidic lumen, the presence of lamp1 and rab7[49-51] but not of early endosomal markers. The current model to explain the mechanism by which VacA induces vacuole formation is by forming anion-selective channels leading to swelling of endosomal compartments.[52-54] What remains unclear is why channel formation would preferentially occur in late endosomes, especially since vacA was also shown to form channels in the plasma

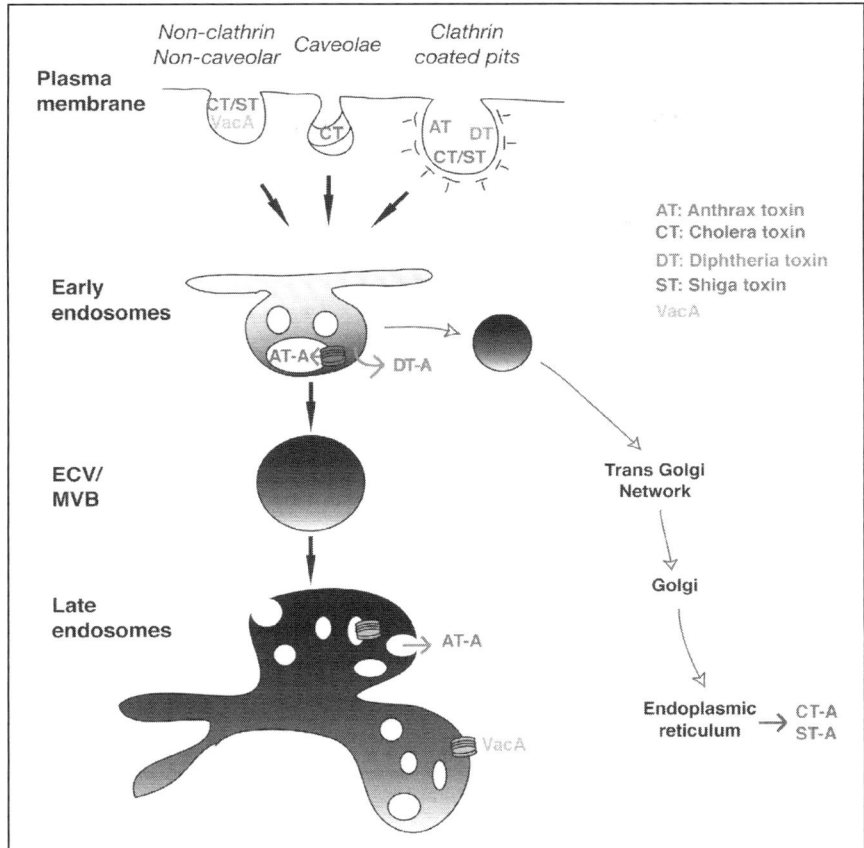

Figure 1. Schematic representation of the intracellular routes followed by anthrax toxin (AT), cholera toxin (CT), Diphtheria toxin (DT), Shiga toxin (ST) and VacA. Toxins can be internalized by a variety of pathways including the clathrin dependent, the caveolar and the clathrin and caveolin independent routes. They are subsequently transported to early endosomes which is either the site from where translocation into the cytoplasm occurs (as for DT) due to membrane insertion of the B subunit at the acidic endosomal pH. Other toxins such as CT and ST are sorted, using protein and lipid based mechanisms, to the trans Golgi Network for subsequent transport to the endoplasmic reticulum where retrotranslocation of the enzymatic A subunits occurs. Yet other toxins such as vacA and anthrax toxin are transported from early endosomes to late endosomes. In this late compartment, vacA forms transmembrane channels thus leading to alterations of the compartment. In contrast anthrax toxin does not alter the compartment but uses it as a entry site to the cytoplasm.

membrane, and in mitochondria.[4] This raises the possibility that channel formation may also occur in early endosomes but does not trigger swelling of the compartment. Note that VacA induced vacuolation only occurs in the presence of weak bases such as ammonium chloride. In the absence of weak bases, despite the absence of vacuolation, alterations of late endosomes where however observed such as inhibition in EGF degradation and procathepsin D maturation[55] or clustering of late endosomes.[51]

In contrast to vacA, anthrax toxin does not alter late endosomes but uses them as a portal of entry into the cytoplasm. As mentioned above, the enzymatic subunits of the anthrax toxin, the lethal factor LF and the edema factor EF, are translocated, at the early endosomal level across the channel formed by the protective antigen from the lumen of the endosomes to the lumen of intraluminal vesicles (Fig. 1). These are subsequently incorporated into endosomal carrier vesicles (ECV), i.e., the transport intermediate that mediate trafficking between early and late endosomes, which are transported along microtubules to the perinuclear regions of the cells where fusion with late endosomes can occur. The requirement for transport of EF and LF to late endosomes is supported by the fact that depolymerization of microtubules using nocodazole or expression of dominant negative rab 7 inhibit the action of LF.[41] Once arrived in late endosomes, EF and LF are thought to await back fusion events between the intraluminal vesicles and the limiting membranes, an event that would lead to release into the cytoplasm. Back fusion between intraluminal vesicles and limiting membrane is a very poorly characterized phenomenon. The occurrence of such events is however supported by the observations that certain proteins found on intraluminal vesicles are merely there on transit. These include the mannose-6-phosphate receptor that is subsequently routed to the Golgi, to capture newly synthesized lysosomal enzymes, as well as major histocompatibility complex II that must reach the plasma membrane once peptide loading has successfully taken place.[56,57] Although the requirement for back fusion has not yet been shown for the anthrax toxin, due to the lack of knowledge concerning this mechanism, the importance of dynamics of intraluminal membranes is supported by the inhibitory effect of antibodies directed towards a lipid almost exclusively found on internal membranes, namely lysobisphosphatidic acid (see Chapter 2). Down regulation of the protein Alix, the mammalian homologue of the yeast class E vacuolar protein sorting vps31, involved in multivesicular body sorting and biogenesis[58,59] was also found to inhibit delivery of LF to the cytoplasm.[41] The benefit of delivering the enzymatic subunits of anthrax toxin to the cytosol from late rather than early endosomes has not yet been demonstrated. It is however attractive to believe that late endosomes act as a delivery platform of these subunits to the vicinity of their early targets.

Concluding Remarks

Bacterial toxins are the fruit of a long-term coevolution between pathogenic bacteria and their host. Trial and error, or attack and counter-attack, have shaped them into very sophisticated weapons that utilize host cell machineries in their greatest intimacy. More over different toxins have developed different strategies in particular to reach the cytoplasmic milieu. Advances in cell biology have considerably helped to better understand the modes of actions of toxins, but importantly the studies of toxins have brought to light unknown or poorly understood intracellular trafficking routes. There is no doubt that these undesired foreigners still have a lot to tell us and will contribute to further characterization of the mechanisms that govern endocytic trafficking.

Acknowledgments

This work was supported by grants by the Swiss National Science Foundation and the National Institute of Health to G. van der Goot. Núria Reig is a recipient of a FEBS fellowship.

References

1. Lencer WI, Hirst TR, Holmes RK. Membrane traffic and the cellular uptake of cholera toxin. Biochim Biophys Acta 1999; 1450(3):177-190.
2. Sandvig K, Grimmer S, Lauvrak SU et al. Pathways followed by ricin and Shiga toxin into cells. Histochem Cell Biol 2002; 117(2):131-141.
3. Collier RJ. Understanding the mode of action of diphtheria toxin: A perspective on progress during the 20th century. Toxicon 2001; 39(11):1793-1803.
4. Cover TL, Blanke SR. Helicobacter pylori VacA, a paradigm for toxin multifunctionality. Nat Rev Microbiol 2005; 3(4):320-332.
5. Naglich JG, Metherall JE, Russell DW et al. Expression cloning of a diphtheria toxin receptor: Identity with a heparin-binding EGF-like growth factor precursor. Cell 1992; 69(6):1051-1061.
6. Bradley KA, Mogridge J, Mourez M et al. Identification of the cellular receptor for anthrax toxin. Nature 2001; 414(6860):225-229.

7. Scobie HM, Rainey GJ, Bradley KA et al. Human capillary morphogenesis protein 2 functions as an anthrax toxin receptor. Proc Natl Acad Sci USA 2003; 100(9):5170-5174.
8. Fujikawa A, Shirasaka D, Yamamoto S et al. Mice deficient in protein tyrosine phosphatase receptor type Z are resistant to gastric ulcer induction by VacA of Helicobacter pylori. Nat Genet 2003; 33(3):375-381.
9. Abrami L, Reig N, van der Goot FG. Anthrax toxin: The long and winding road that leads to the kill. Trends Microbiol 2005; 13(2):72-78.
10. Scobie HM, Young JA. Interactions between anthrax toxin receptors and protective antigen. Curr Opin Microbiol 2005; 8(1):106-112.
11. Abrami L, Liu S, Cosson P et al. Anthrax toxin triggers endocytosis of its receptor via a lipid raft-mediated clathrin-dependent process. J Cell Biol 2003; 160(3):321-328.
12. Wolf AA, Jobling MG, Wimer-Mackin S et al. Ganglioside structure dictates signal transduction by cholera toxin and association with caveolae-like membrane domains in polarized epithelia. J Cell Biol 1998; 141(4):917-927.
13. Schraw W, Li Y, McClain MS et al. Association of helicobacter pylori vacuolating toxin (VacA) with lipid rafts. J Biol Chem 2002; 277:34642-34650.
14. Spilsberg B, Hanada K, Sandvig K. Diphtheria toxin translocation across cellular membranes is regulated by sphingolipids. Biochem Biophys Res Commun 2005; 329(2):465-473.
15. van der Goot FG, Gruenberg J. Oiling the wheels of the endocytic pathway. Trends Cell Biol 2002; 12(7):296-299.
16. Conner SD, Schmid SL. Regulated portals of entry into the cell. Nature 2003; 422(6927):37-44.
17. Parton RG, Richards AA. Lipid rafts and caveolae as portals for endocytosis: New insights and common mechanisms. Traffic 2003; 4(11):724-738.
18. Ricci V, Galmiche A, Doye A et al. High cell sensitivity to helicobacter pylori VacA toxin depends on a GPI-anchored protein and is not blocked by inhibition of the clathrin- mediated pathway of endocytosis. Mol Biol Cell 2000; 11(11):3897-3909.
19. Moya M, Dautry-Varsat A, Goud B et al. Inhibition of coated pit formation in Hep2 cells blocks the cytotoxicity of diphtheria toxin but not that of ricin toxin. J Cell Biol 1985; 101(2):548-559.
20. Boll W, Ehrlich M, Collier RJ et al. Effects of dynamin inactivation on pathways of anthrax toxin uptake. Eur J Cell Biol 2004; 83(6):281-288.
21. Sandvig K, van Deurs B. Transport of protein toxins into cells: Pathways used by ricin, cholera toxin and Shiga toxin. FEBS Lett 2002; 529(1):49-53.
22. Orlandi PA, Fishman PH. Filipin-dependent inhibition of cholera toxin: Evidence for toxin internalization and activation through caveolae-like domains. J Cell Biol 1998; 141(4):905-915.
23. Torgersen ML, Skretting G, van Deurs B et al. Internalization of cholera toxin by different endocytic mechanisms. J Cell Sci 2001; 114(Pt 20):3737-3747.
24. Shogomori H, Futerman AH. Cholera toxin is found in detergent-insoluble rafts/domains at the cell surface of hippocampal neurons but is internalized via a raft-independent mechanism. J Biol Chem 2001; 276(12):9182-9188.
25. Massol RH, Larsen JE, Fujinaga Y et al. Cholera toxin toxicity does not require functional Arf6- and dynamin-dependent endocytic pathways. Mol Biol Cell 2004; 15(8):3631-3641.
26. Hansen GH, Dalskov SM, Rasmussen CR et al. Cholera toxin entry into pig enterocytes occurs via a lipid raft- and clathrin-dependent mechanism. Biochemistry 2005; 44(3):873-882.
27. Katagiri YU, Mori T, Nakajima H et al. Activation of Src family kinase yes induced by Shiga toxin binding to globotriaosyl ceramide (Gb3/CD77) in low density, detergent-insoluble microdomains. J Biol Chem 1999; 274(49):35278-35282.
28. Mori T, Kiyokawa N, Katagiri YU et al. Globotriaosyl ceramide (CD77/Gb3) in the glycolipid-enriched membrane domain participates in B-cell receptor-mediated apoptosis by regulating lyn kinase activity in human B cells. Exp Hematol 2000; 28(11):1260-1268.
29. Geisse NA, Cover TL, Henderson RM et al. Targeting of Helicobacter pylori vacuolating toxin to lipid raft membrane domains analysed by atomic force microscopy. Biochem J 2004; 381(Pt 3):911-917.
30. Gauthier NC, Ricci V, Gounon P et al. Glycosylphosphatidylinositol-anchored proteins and actin cytoskeleton modulate chloride transport by channels formed by the Helicobacter pylori vacuolating cytotoxin VacA in HeLa cells. J Biol Chem 2004; 279(10):9481-9489.
31. Parton RG. Ultrastructural localization of gangliosides; GM1 is concentrated in caveolae. J Histochem Cytochem 1994; 42(2):155-166.
32. Nichols BJ. A distinct class of endosome mediates clathrin-independent endocytosis to the Golgi complex. Nat Cell Biol 2002; 4(5):374-378.
33. Pelkmans L, Burli T, Zerial M et al. Caveolin-stabilized membrane domains as multifunctional transport and sorting devices in endocytic membrane traffic. Cell 2004; 118(6):767-780.

34. Kirkham M, Fujita A, Chadda R et al. Ultrastructural identification of uncoated caveolin-independent early endocytic vehicles. J Cell Biol 2005; 168(3):465-476.
35. Sabhararanjak S, Sharma P, Parton RG et al. GPI-anchored proteins are delivered to recycling endosomes via a distinct cdc42-regulated, clathrin-independent pinocytic pathway. Dev Cell 2002; 2:411-423.
36. Saint-Pol A, Yelamos B, Amessou M et al. Clathrin adaptor epsinR is required for retrograde sorting on early endosomal membranes. Dev Cell 2004; 6(4):525-538.
37. Lauvrak SU, Torgersen ML, Sandvig K. Efficient endosome-to-Golgi transport of Shiga toxin is dependent on dynamin and clathrin. J Cell Sci 2004; 117(Pt 11):2321-2331.
38. Gruenberg J. The endocytic pathway: A mosaic of domains. Nat Rev Mol Cell Biol 2001; 2(10):721-730.
39. Miaczynska M, Zerial M. Mosaic organization of the endocytic pathway. Exp Cell Res 2002; 272(1):8-14.
40. Lemichez E, Bomsel M, Devilliers G et al. Membrane translocation of diphtheria toxin fragment A exploits early to late endosome trafficking machinery. Mol Microbiol 1997; 23(3):445-457.
41. Abrami L, Lindsay M, Parton RG et al. Membrane insertion of anthrax protective antigen and cytoplasmic delivery of lethal factor occur at different stages of the endocytic pathway. J Cell Biol 2004; 166:645–651.
42. Ratts R, Zeng H, Berg EA et al. The cytosolic entry of diphtheria toxin catalytic domain requires a host cell cytosolic translocation factor complex. J Cell Biol 2003; 160(7):1139-1150.
43. Fujinaga Y, Wolf AA, Rodighiero C et al. Gangliosides that associate with lipid rafts mediate transport of cholera and related toxins from the plasma membrane to ER. Mol Biol Cell 2003.
44. Falguieres T, Mallard F, Baron C et al. Targeting of Shiga toxin B-subunit to retrograde transport route in association with detergent-resistant membranes. Mol Biol Cell 2001; 12(8):2453-2468.
45. Wilcke M, Johannes L, Galli T et al. Rab11 regulates the compartmentalization of early endosomes required for efficient transport from early endosomes to the trans-golgi network. J Cell Biol 2000; 151(6):1207-1220.
46. Tai G, Lu L, Wang TL et al. Participation of the syntaxin 5/Ykt6/GS28/GS15 SNARE complex in transport from the early/recycling endosome to the trans-Golgi network. Mol Biol Cell 2004; 15(9):4011-4022.
47. Mallard F, Antony C, Tenza D et al. Direct pathway from early/recycling endosomes to the Golgi apparatus revealed through the study of shiga toxin B-fragment transport. J Cell Biol 1998; 143(4):973-990.
48. Lencer WI, Tsai B. The intracellular voyage of cholera toxin: Going retro. Trends Biochem Sci 2003; 28(12):639-645.
49. Papini E, Satin B, Bucci C et al. The small GTP binding protein rab7 is essential for cellular vacuolation induced by Helicobacter pylori cytotoxin. EMBO J 1997; 16(1):15-24.
50. Morbiato L, Tombola F, Campello S et al. Vacuolation induced by VacA toxin of Helicobacter pylori requires the intracellular accumulation of membrane permeant bases, Cl(-) and water. FEBS Lett 2001; 508(3):479-483.
51. Li Y, Wandinger-Ness A, Goldenring JR et al. Clustering and redistribution of late endocytic compartments in response to Helicobacter pylori vacuolating toxin. Mol Biol Cell 2004; 15(4):1946-1959.
52. Czajkowsky DM, Iwamoto H, Cover TL et al. The vacuolating toxin from Helicobacter pylori forms hexameric pores in lipid bilayers at low pH. Proc Natl Acad Sci USA 1999; 96(5):2001-2006.
53. Tombola F, Carlesso C, Szabo I et al. Helicobacter pylori vacuolating toxin forms anion-selective channels in planar lipid bilayers: Possible implications for the mechanism of cellular vacuolation. Biophys J 1999; 76(3):1401-1409.
54. Szabo I, Brutsche S, Tombola F et al. Formation of anion-selective channels in the cell plasma membrane by the toxin VacA of Helicobacter pylori is required for its biological activity. EMBO J 1999; 18(20):5517-5527.
55. Satin B, Norais N, Telford J et al. Effect of helicobacter pylori vacuolating toxin on maturation and extracellular release of procathepsin D and on epidermal growth factor degradation. J Biol Chem 1997; 272(40):25022-25028.
56. Kobayashi T, Vischer UM, Rosnoblet C et al. The tetraspanin CD63/lamp3 cycles between endocytic and secretory compartments in human endothelial cells. Mol Biol Cell 2000; 11(5):1829-1843.
57. Chow A, Toomre D, Garrett W et al. Dendritic cell maturation triggers retrograde MHC class II transport from lysosomes to the plasma membrane. Nature 2002; 418(6901):988-994.
58. Odorizzi G, Katzmann DJ, Babst M et al. Bro1 is an endosome-associated protein that functions in the MVB pathway in Saccharomyces cerevisiae. J Cell Sci 2003; 116(Pt 10):1893-1903.
59. Strack B, Calistri A, Craig S et al. AIP1/ALIX is a binding partner for HIV-1 p6 and EIAV p9 functioning in virus budding. Cell 2003; 114(6):689-699.

Index